CONSTRUCTING BUILDING ENCLOSURES

Constructing Building Enclosures investigates and interrogates tensions that arose between the disciplines of architecture and engineering as they wrestled with technology and building cultures that evolved to deliver structures in the modern era. At the center of this history are inventive architects, engineers and projects that did not settle for conventional solutions, technologies and methods.

Comprised of thirteen original essays by interdisciplinary scholars, this collection offers a critical look at the development and purpose of building technology within a design framework. Through two distinct sections, the contributions first challenge notions of the boundaries between architecture, engineering and construction. The authors then investigate twentieth-century building projects, exploring technological and aesthetic boundaries of postwar modernism and uncovering lessons relevant to enclosure design that are typically overlooked. Projects include Louis Kahn's Weiss House, Minoru Yamasaki's Science Center, Sigurd Lewerentz's Chapel of Hope and more.

An important read for students, educators and researchers within architectural history, construction history, building technology and design, this volume sets out to disrupt common assumptions of how we understand this history.

Clifton Fordham is a registered architect and Assistant Professor at Temple University, where he teaches building design and building technology. His current focus is building enclosures with an emphasis on how their design relates to the sun. He is a graduate of the Yale School of Architecture and Howard University.

CONSTRUCTING BUILDING ENCLOSURES

Architectural History, Technology and Poetics in the Postwar Era

Edited by
Clifton Fordham

Routledge
Taylor & Francis Group

NEW YORK AND LONDON

First published 2021
by Routledge
52 Vanderbilt Avenue, New York, NY 10017

and by Routledge
2 Park Square, Milton Park, Abingdon, Oxon, OX14 4RN

Routledge is an imprint of the Taylor & Francis Group, an informa business

© 2021 Taylor & Francis

Library of Congress Cataloging-in-Publication Data
A catalog record for this book has been requested

ISBN: [978-0-367-27628-7] (hbk)
ISBN: [978-0-367-27625-6] (pbk)
ISBN: [978-0-429-29696-3] (ebk)

Typeset in Bembo
by Apex CoVantage, LLC

CONTENTS

NOTES ON CONTRIBUTORS

The following are biographies of contributors to this book, listed in alphabetical order:

Mary Ben Bonham, Associate Professor at Miami University in Oxford, Ohio, received a BArch from the University of Texas at Austin and a MArch from the University of Pennsylvania. She teaches and researches in the areas of building design, technology and sustainability with an emphasis on lighting and facades. Bonham is the author of *Bioclimatic Double-Skin Facades* (Routledge, 2019), a book that examines how double-skin facades have been adapted to a range of climates and cultural settings.

Andrew Cruse is Associate Professor of Architecture at The Ohio State University and a registered architect. His academic work focuses on the evolution of human comfort and its impact on architectural design. His design practice, Good Form Studio, recently completed a project for Novartis Pharma in Basel, Switzerland. Prior to joining the faculty at Ohio State, Cruse taught at Washington University in St. Louis and was an associate at Machado and Silvetti in Boston. He holds a BA in Art History from Columbia University and an MArch from Rice University.

Mahyar Hadighi is an architectural designer, historic preservationist and educator. He is Assistant Professor of Architecture and Director of the Historic Preservation and Design program at Texas Tech University. As an architect and a historic preservationist, Mahyar concentrates on modernism through his work of documenting and analyzing local adaptations of modern architecture using computational design methodologies.

Matthew Hall, is an architect and Associate Professor at the Auburn University College of Architecture, Design and Construction and Director of the Scandinavian Study Abroad Program. He is principal and designer at the multidisciplinary design firm Obstructures with a focus on the impact of material culture and the problematic nature of design. Current research is concentrated on Swedish architects Sigurd Lewerentz and Bernt Nyberg. In 2017, Hall served as guest editor for a special issue on the work of Nyberg for A+U (*A+U 564 17:09*); he has lectured, published and exhibited internationally on these and other topics related to theory, criticism and pedagogy.

Tait Johnson is an architect and architectural historian teaching at the University of Illinois at Urbana-Champaign. Johnson researches the history and theory of modern architecture and materiality in the twentieth century, specifically concerned with the image and instrumentality of aluminum in architecture. His research has been published in the *International Journal of the Constructed Environment* and presented at the Vernacular Architecture Forum, the Society of Architectural Historians annual conference and meetings of the Construction History Society of America. Ongoing research explores the history of speculative design and futurist architecture.

Thomas Leslie, FAIA, is the Morrill Professor in Architecture at Iowa State University, where he teaches building science, design and history. He is the author of *Beauty's Rigor: Patterns of Production in the Work of Pier Luigi Nervi* and *Chicago Skyscrapers, 1871–1934*.

Whitney Moon is Assistant Professor of Architecture at University of Wisconsin-Milwaukee, where she teaches history, theory and design. Her research interests reside in twentieth- and twenty-first-century art and architecture, with an emphasis on theatricality, performance and ephemeral works. Her writings on pneumatics have been published in *PRAXIS*, *e-flux*, *JAE*, *Room One Thousand*, *The Other Architect* and *Dialectic*. A registered architect in California and Wisconsin, Moon earned her PhD in Architectural History and Theory from University of California, Los Angeles, and BArch from California Polytechnic State University, San Luis Obispo.

Scott Murray is a licensed architect and Associate Professor of Architecture at the University of Illinois at Urbana-Champaign. His research focuses on the cultural significance of architectural technology of the last 100 years, with an emphasis on innovative building-envelope design. He is the author of the books *Contemporary Curtain Wall Architecture* (Princeton Architectural Press) and *Translucent Building Skins* (Routledge) and serves on the Editorial Board of *Technology | Architecture + Design*, an international peer-reviewed journal published by the Association of Collegiate Schools of Architecture.

Ute Poerschke is Professor at the Pennsylvania State University. Her research focuses on the relationship of architecture and technology, the theory of functionalism and the integrative architectural design process. She is a licensed architect and urban planner in Germany and a principal of the firm Friedrich-Poerschke-Zwink Architects and Urban Planners in Munich, Germany.

Meredith Sattler is Assistant Professor of Architecture at Cal Poly San Luis Obispo, a PhD candidate in Science and Technology Studies at Virginia Tech and a LEED BD+C. She received her Master of Architecture and Master of Environmental Management degrees from Yale University, and her Bachelor of Arts from Vassar College. Her interests include conceptualizations of dynamic sustainable constructed environments; interdisciplinary structure and practices between designers, engineers and scientists; and design's agency within natural environment–technology–human interactions.

Joseph M. Siry, William R. Kenan, Jr., Professor of the Humanities, Professor of Art History, Wesleyan University, received an AB in History from Princeton University, an MArch from the University of Pennsylvania and a PhD in history, theory, and criticism of architecture from MIT. His books are *Carson Pirie Scott: Louis Sullivan and the Chicago Department Store* (Chicago, 1988); *Unity Temple: Frank Lloyd Wright and Architecture for Liberal Religion* (Cambridge, 1996); *The Chicago Auditorium Building: Adler and Sullivan's Architecture and the City* (Chicago, 2002); and *Beth Sholom Synagogue: Frank Lloyd Wright and Modern Religious Architecture* (Chicago, 2012). His book *Air-Conditioning and Modern American Architecture, 1890–1970: From Adler and Sullivan to Louis Kahn*, is forthcoming from Pennsylvania State University Press.

Tyler S. Sprague is Assistant Professor in the Department of Architecture at the University of Washington. He has a background in structural engineering and a PhD in architectural history. His book, *Sculpture on a Grand Scale: Jack Christiansen's Thin Shell Modernism*, was published by the University of Washington Press in 2019.

Rob Whitehead, AIA, LEED AP, is a licensed architect and Associate Professor of Architecture at Iowa State University, where he teaches structural design, integrated design studios and design-build. His work is centered on the conjunction of architecture, structural design and construction in pedagogy and practice. He is the author of *Structures by Design: Thinking, Making, Breaking* (Routledge, 2019) and co-author of *Design-Tech: Building Technology for Architects* (Routledge, 2014) with Tom Leslie and Jason Alread.

ACKNOWLEDGMENTS

Although it is not feasible to acknowledge all who were critical to the project, I must recognize certain individuals. First, I am grateful to Krystal LaDuc at Routledge for seeing value in this project. I thank Dean Susan Cahan, Associate Dean Kate Wingert-Playdon and Architecture Department Chair Rashida Ng at the Tyler School of Art and Architecture for their support and patience. As an editor and author, I could count on Stephen Anderson, Bill Craig and Seher Erdogan Ford to provide candid feedback. I also thank Marty Henry for his candid advice and enthusiasm, which has been invaluable.

Without the trailblazers at the Construction History Society of America, especially Brian Bowen, an audience for this type of work in the United States would be less fertile. Thomas Leslie introduced me to the organization and has provided valuable guidance. I'm grateful to William Whitaker and Heather Isabel-Schumacher at the University of Pennsylvania Architectural Archives for allowing me to fumble through many valuable documents in the collection and for providing important insight into its content. Although it seems like an eternity ago, I am grateful to Patrick Pinnell, who coached me during his seminars at the Yale School of Architecture. His patience and passion have continued to influence my teaching. Finally, I thank Joanne for tolerating work that kept me absorbed for what seems like an eternity, and my mother for encouraging me to pursue my passions.

INTRODUCTION

Enclosure Expanded

Clifton Fordham

Enclosures are the most important part of a building both functionally and visually. They lie at the intersection of environmental factors and elements that impact buildings: structure, thermal, wind, moisture, acoustics, light, privacy, air quality, security, durability, ease of assembly, regulatory, economic, adjustability, adaptability and so on. The relative importance of these factors varies across project type and is actively addressed in varying degrees. Some issues, such as damage or discomfort from concentrated light reflected off convex glass curtain walls, would not have been imaginable a few decades ago. Similarly, condensation inside exterior wall assemblies is a problem of tightly sealed buildings, which themselves are a result of enclosure expectations related to more exacting climate control of building interiors after World War II. Complicating effective performance of building enclosures systems are interlocking dependencies from the intersection of different materials systems and design criteria. For example, structural systems expand at different rates under thermal stress; acoustic and thermal insulation are not symmetrical; different amounts and location of glazing impact mechanical systems and artificial lighting; and flashing strategies impact the aesthetics of buildings.

With all the complexity inherent in building enclosures, it is peculiar that their appearance has gained almost all of the attention provided to them in historical accounts, while underlying construction realities remain almost entirely unaddressed. Perhaps the notion that building enclosures are seen by more people than building interiors warrants an almost exclusively visual account of exteriors since the messages that buildings convey to the public are very important. Nonetheless, there are pitfalls inherent in overlooking technical realities that are inescapable in the life of a building and demand significant capabilities to design and realize.

A culture of architectural production that privileges and separates the look of buildings from their technical aspects, often perceived as a necessary burden to bear, is a corollary problem. Within this paradigm, performance issues are equated with liability, and responsibility for their execution is relegated to technical architects, consultants and fabricators, who are largely invisible in historical and contemporary accounts. Separating design and production also reinforces a hierarchy where visual results are privileged over technical results unless the technical results are emphasized in the visual expression of surfaces. This contributes to a loss of appreciation for nuanced details related to the production of building enclosures and a layered understanding of history. Aesthetic autonomy relieves architects of some accountability but does not eliminate the reality that aesthetics and technical performance are interdependent, not exclusive.

This volume brings together essays that address the integration of both aesthetic and technical realities of building enclosures, focusing on details and how they are constructed. It centers on modern building enclosures built in the period following World War II, when optimism from the mobilization of technology toward industry resonated most acutely in America, in a burst of pent-up architectural production. The result was a maturing of the modernist impulse mated with technological capability that better reflected the aspirations of prewar masters. A postwar context revealed synergies and tensions between architecture, engineering, construction and industry that produced work more nuanced and interesting than is generally acknowledged. The technology of postwar buildings also offers valuable, accessible, and applicable lessons for a contemporary era that favors aesthetic and technical complexity to a degree unimaginable in the past. This book juxtaposes issues related to building production and the material of building technology, reflected in the fertile ground of the construction history field, in a manner more natural to engineers and builders than architects. Greater focus on building activity is also coupled with a more nuanced and comprehensive look at detailed components of building assemblies critical to fabrication. Unlike most construction histories, the building enclosure, the most dynamic building system and locus for expression, is the central subject.

Problems of Exclusion

Le Corbusier's Villa Savoye provides an example of how historical analysis has primarily been aesthetic, separated from realities of the building that impacted the building's failure to meet expectations as a house. Like buildings of the same period, it was single glazed, but the relative proportion of glass to radiant heating elements critical for comfort were poorly coordinated, resulting in cold spaces.[1] This raises the question of whether the formal goals of the project could have been achieved without this sensory liability. In *Allure of the Incomplete, Imperfect and Impermanent*, Rumiko Handa provides a non-flattering history of Corbusier's structure that highlights exterior wall finishes that prematurely aged. She presents this case as part of a larger critique of architectural history that fixes buildings in time, by omitting the aging of buildings, thus limiting an authoritative perspective of how they truly perform.[2] Similarly, in *On Weathering: The Life of Buildings in Time*, Moshen Mostafavi and David Leatherbarrow explored the role played by wear from natural elements, challenging the paradigm in which modern architecture is recorded as fixed in time, continually presented as new. Images used in the book are considerably less flattering representations of structures than those normally recorded in history books. This contrasts with how Le Corbusier saw weathering of his 1920s buildings as a departure from their ideal state of new.[3]

Like freezing architecture in time or omitting material details, narratives that exclude the construction of architecture, how it is made and the properties of materials, come with complications and risks. Construction is contextual, dependent on materials, environment, economics, tools and people. It is also logistical, and the results are often messy when not finished or completed, appearing non-rational despite pragmatic objectives and methods. It is not surprising that engineers have out-represented architects as builders since the late nineteenth century, and that practical historical accounts of building construction are currently more likely to come from an engineering perspective than an architectural one. In contrast to the more expressive parts of architecture, architects as a group have shied away from including construction details, technical documentation, building-site photos and process narratives in their historical accounts. This mirrors an aversion by some contemporary architects to technical details of design and activities on the construction site.

It is also important to recognize methods by which buildings are erected that provide cultural and technological context, offering insight into conditions that shape buildings. Methods of building construction ground the applicability of technologies and aesthetics, allowing architects to administer designs appropriate to time and place. This raises another contradiction in the Villa Savoye that has contemporary corollaries: the alignment of design visions with realistic construction cultures. The erection of the Villa Savoye was more medieval than modern, belying the technological progress that the building expressed. Materials on the surfaces of its construction site were distributed haphazardly. Rough scaffolding provided a platform by which labor proceeded slowly and inconsistently without machines (Figure I.1). The unit masonry below the final plaster finishes were rough and inconsistent.[4] A more accurate understanding of how the Villa Savoye was built calls into question the alignment of technologies and methods with the idea of what architecture should be, what it should represent and what kind of solutions designers offer.[5]

In the case of the Villa Savoye, correlation between technology and design is significant because the intention of representing industrialization with a

FIGURE I.1   Photograph of the Villa Savoye under construction, 1929.

Source: Archives of the Foundation Le Corbusier/ADAGP.

new typology created new burdens for architecture. It is the subject of representation, over the material realities of the building and its construction, that has been privileged by architectural historians. Few historical accounts have acknowledged inconsistencies between the construction of the Villa Savoye and theories propagated by Le Corbusier prior to its construction that lend the building much of its significance. In his book *Towards a New Architecture*, Corbusier extolled the virtues of engineering visible in aircraft and ships and modern manufacturing processes exemplified in the assembly line of the Model T.[6] This contradicts the material reality of the Villa Savoye, which was constructed on site with few components assembled in a factory.[7] Similarly, photographs of the Villa Savoye during construction demonstrate contrasts between stated ideals and elements of construction that did not contribute to his larger aesthetic visions. This mirrored Corbusier's habit of photographing projects under construction from the same perspective as his photographs of completed works, omitting perspectives important to comprehending building details.[8] Without an account of the construction and underlining technology of buildings, ideas prevail over reality, tempting architects to discount responsibility for integrated design and construction.

And yet, acknowledging a fuller scope of building and underlying design intentions creates a risk of diluting the pool of available discourse that aids in advancing contemporary design efforts. Perhaps this dilution discourages big creative leaps and saps appreciation of contributions such as Corbusier's five points of architecture. Another perspective is that those contributions will persevere, and that alternative perspectives can concurrently be appreciated that involve a closer look at how buildings are realized where larger concepts are not pure.

The reality of construction, however, is that it involves compromise, group intelligence and skills that extend beyond design conception but do not exclude it. This is consistent throughout the history of architecture, but did not become a defining issue until the early twentieth century, when technological complexity and project delivery expectations called for specialization. By the time Reyner Banham wrote his seminal book, *The Architecture of the Well-Tempered Environment*, first published in 1969, the disciplines of architecture and engineering had formally established their distinct roles in the development of buildings. Although architects were technical, the activities of engineers were clearly more technical.[9] Specialization went hand in hand with the increased cost of building systems, expectations that the systems perform predictably and demands of owners who increasingly looked to engineers for guidance. Despite the importance of engineered systems and building enclosures, because large portions of those systems are often hidden their inclusion in architectural histories remains limited.

Building Technology and Architectural History

The paucity of technological analysis by architectural historians is too often taken for granted or dismissed as inconsequential. Lack of acknowledgment paralleled a tenuous position toward technology in general on the part of leading architectural historians at the time *The Architecture of the Well-Tempered Environment* was written, a reality that still lingers in nuanced ways. The general relationship of building technology to architectural history was the subject of Banham's introduction to the second edition of the book he calls *Unwarranted Apology*. He writes:

> The idea that architecture belongs in one place and technology in another is comparatively new in history, and its effect in architecture, which should be the most complete of the arts of mankind, has been crippling. In the eighteenth century, at least as late as Isaac Ware's *Complete Body of Architecture* (1756), that body had indeed been complete, and the technology then available had found a comfortable place within its compendious pages. Thereafter, however, the art of architecture became increasingly divorced from the practice of making and operating buildings.[10]

Banham later describes difficulties that leaders of his discipline had fitting his book into the traditional expectations of architecture history due to the central role of comfort systems in his analysis. His critics went as far as to say that the topic had been exhausted, referencing Sigfried Giedion's book *Mechanization Takes Command* as conclusive, leaving little room for continued historical accounts of technology.[11]

Despite the scarcity of technology in architectural histories when Banham wrote his book, technological dimensions of building design were accounted for in architectural journals and technical manuals. This raises questions as to the significance of compartmentalization and the trajectory of architectural practice relative to engineering disciplines that continued to exercise influence over large portions of building design and budgets. For Banham, the omissions were important because of a perceived disjuncture between technology of the time and architectural expression that located architects behind society, rearguard rather than avant-garde. Pop-artists who acknowledged the populist influence of mass marketing, throwaway consumerism and techno-euphoria associated with the military–industrial complex embraced these larger cultural trends. Not embracing these changes risked cultural irrelevancy.[12]

Impetus for expanding the scope of architecture and scholarly work that related to enclosure technology primarily came from the growth of the environmental and preservation movements. Examples of historic preservation in the United States date back to the early 1800s in successful attempts to preserve Independence Hall in 1816 and Mount Vernon in 1853. Other early actions included the restoration of Colonial Williamsburg, which started in 1926.[13] The practical activity of assessing and restoring older buildings necessitated an understanding of building technology from a historical perspective. A disjuncture between current codes and material technology made restoring old enclosures difficult. For example, the lime content in mortar was higher a hundred years ago than today. Using contemporary Portland cement-based mortar mixes damages older bricks, which are softer. The Association for Preservation Technology was founded in 1966 to address such problems.

Increased awareness of pollution, energy scarcity and the relationship of the built and natural environment to individual well-being focused many in the architectural community upon the non-visual impacts of building performance. Included in this activity were reprints in the 1970s of Victor Olgyay's 1963 book *Design With Climate*. Many built examples of environmental architecture were folksy, incorporating technologies such as water tanks located inside windows modeled on principles gleaned from studying vernacular examples of thermal masses. With few exceptions, most of the architectural projects inspired by the environmental movement and corresponding academic research did not jell with dominant cultural narratives, but over time gained enough acceptance to stand on their own. A more current academic corollary to Banham's *Well-Tempered Environment* is *The Environmental Imagination: Technics and Poetics of the Architectural*

Environment by Dean Hawkes. Published in 2008, it balances technical and aesthetic analysis, contextualizing historic building in contemporary environmental science.

Construction History

While building technology manuals, bulletins and textbooks provide a foundation for architectural knowledge, scholarly non-historical work on building technology is primarily written for technically inclined architects, engineers and building scientists. On the other hand, histories of building technology since *The Architecture of the Well-Tempered Environment* have mostly been written by individuals who do not have a background as architectural historians or technically inclined architects. The core field for building technology history has come to be known as construction history and is widely understood as a branch of the history of technology with objectives that distinguish it from the latter group. Its vanguard organization is the Construction History Society of Britain, founded in 1982. Similar groups have since formed in France, Germany, Spain and the United States.[14]

In a 1985 journal editorial for the newly founded *Construction History* journal, John Summerson defined construction history as the history of structural design and the history of building practice. He further specifies that structural design in this case is a study of particular materials and speculates that the future audience for the journal is the "building world."[15] Summerson, who was an architectural historian, divided contributors to construction history into two groups: those who study materials and those who study practices and techniques undertaken to fashion those materials into built works. Demonstrating the evolutionary nature of the discipline, as Andrew Saint noted in an article two decades later, the focus of contributions to the journal had shifted heavily toward structural engineering and away from practice, the latter of which motivated the Construction History Society's founders.[16] This is evidenced in the proportion of contributions to the journal that have revolved around a specific material such as wood, iron or concrete.

Concentrated study of material systems, especially antiquated ones that generally address limited technical issues, has come at the expense of aesthetic issues and the relationships between people, raising a question of the relevance of construction history to engineers and architects who solve immediate problems. Saint insightfully points out that the implications of a robust history are less significant for engineers than architects. Engineers and architects who exclusively solve technical problems have limited need for history except prior to a point of action. Architects are not bound to a notion of progress in which solutions need to develop in a linear manner and are inclined to find solutions across the spectrum of time, thus lending themselves to using history as a design aid. However, engineers also diverge from completely rational problem-solving activity, blurring absolute distinctions between technician and artist. Along with a different relationship to history, a distinguishing factor between engineers and architects is, first, the demanding depth of fluency in mathematics and scientific principles required for steel and concrete design that calls for the specialization of engineers and, second, the demands of complex programs and statements worthy of representing institutions, the specialization of architecture.[17]

Specialization has resulted in an operative dependency in which different complementary skills are brought together to accomplish tasks with various levels of complexity and scale. Despite the tremendous size and significance of the building industry, chronicling the complexity of the building process has been a fragmented and incomplete affair. The broader field of technology history has favored the study of materials, systems and industrial process, plus practical industrial products such as machines and automobiles, which are of similar interest to society as both artifacts and participants in larger societal relationships. Building construction, which is site specific and brings together disparate facilitators, is more difficult to study than industrial processes, which are streamlined and better organized, explaining the need for a distinct historical discipline.

Construction and building design are also tangled processes with responsibility stratified among participants, revealing a complex mesh of artistic and technical activity. Within the design process, responsibilities have been assigned to architects and specialist engineers who are relatively independent. This is reflected in how different project contributors convey design information. For example, content generated by structural engineers appear on separate sheets of paper for which they take primary responsibility. With enclosure design, varying performance-oriented requirements are tightly meshed and largely inseparable from issues of form and communication. Despite the technical nature of enclosures, there is a recent trend toward architects farming out the technical design of facades to outside consultancies for larger projects when budgets allow. Architects risk relinquishing a leadership position in enclosure systems, diluting one of the most important roles of the architect where aesthetics and technology can't easily be separated. It is unlikely engineers will fully embrace the artistic part of building design in a manner that satisfies societal desires, needs and expectations.

Enclosure History

For architects, the need for history is more acute than for engineers. This leaves a question of who should write much needed histories. It is questionable if engineers, or others, will develop enough insight and interest in the technical details of building enclosures *and* consider the aesthetic nature of building enclosures in their assessments. Promise lies with recent general construction history research that encompasses building enclosure, which has typically emerged from individuals with a background in architectural or structural engineering practice. This is largely because practice experiences typically draw upon technical knowledge and hone the ability to recognize its relevance to complementary objectives. Similarly, a sub-specialty of construction history is needed that satisfies demand for greater knowledge of issues related to building enclosures.

Construction history books, such as John Fitchen's *Building Construction Before Mechanization*, provide a thorough examination of enclosures built before the nineteenth century, albeit without an aesthetic perspective. Books by Thomas Leslie on the works of Louis Kahn and Pier Luigi Nervi include detailed analysis from both a technology and an aesthetic perspective, and at various scales. And, as mentioned, a limited number of individual construction history essays with similar balance have been distributed in different venues. The most comprehensive investigations of building enclosures in this vein are included in *The Details of Modern Architecture* and *The Details of Modern Architecture II* by Edward Ford, published in 1990 and 1996 respectively. Ford's books juxtapose beautifully drafted illustrations of building details and black and white photographs with extensive narrative. One of the benefits of his axonometric drawings is that they peel away key layers of building enclosures, indicating how materials below the skin of historically significant buildings relate to each other (Figure I.2). Ford's analysis of master architects and their buildings includes narrative descriptions of the detail drawings and photographs that he relates to larger themes in architectural history and materials. What he does not accomplish is relating the physical realties of building enclosures to efforts between builders and project contributors within a socio-economic context. A richer historical account of building enclosure considering these influences has barely been breached and warrants expansion considering the collaborative nature of the enclosure design and delivery processes.

Technology and Details

The reality of architectural practice, engineering and construction changed dramatically between the late nineteenth century and the postwar period. However, the way architectural history is recorded has yet to reflect this shift in a comprehensive way, especially in relation to details and their design. Prior to the 1800s, most of the details for built works would have been worked out on-site and not drafted in advance of construction, as is the case now.[18] When the need for details first occurred, the task of detailing initially fell on fabricators and later became a steady part of architecture office production as assemblies became more complex. An example can be found in the transition from load-bearing masonry enclosure in high-rises, such as with the 17-story high Manadnock Building (1881–91), to structural skins supported by a distinct structural frame (Figure I.3). Structural steel frames allowed for a thin post-and-beam structure but necessitated infill and protection from different materials since steel softens when exposed to high temperatures from fire. The answer to these shortcomings was a combination of plate glass, terracotta and brick masonry found in the exterior details of the Reliance Building (1890–95). Connecting all this material were miscellaneous steel

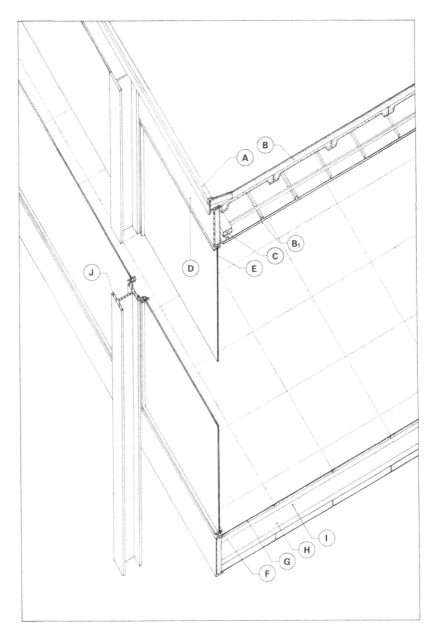

FIGURE I.2 Axonometric. Edward Ford, *The Details of Modern Architecture* (Cambridge: MIT Press, 1990).

Source: MIT Press.

angles and connections that had to be designed and coordinated with cladding formed off-site prior to assembly (Figure I.3). The details and coordination of these assemblies became the responsibility of architecture practices.

The significance of the emergence of the architectural detail on the practice of architecture has not been sufficiently recognized and the ramifications of corresponding responsibilities are unresolved. Detailing created a new technological role within architectural practices, greatly expanding the amount of time and effort necessary to design a building. Large-scale conceptualization of buildings was further concentrated within a smaller fraction of firm leaders. The distance between conception and realization was compounded by the increased technical nature of the coordination of building systems, which were less forgiving than before, and the knowledge and skills necessary to document and detail. While the majority of firm leaders still benefited from higher education, experience gained on the construction site was the historical path for most junior firm

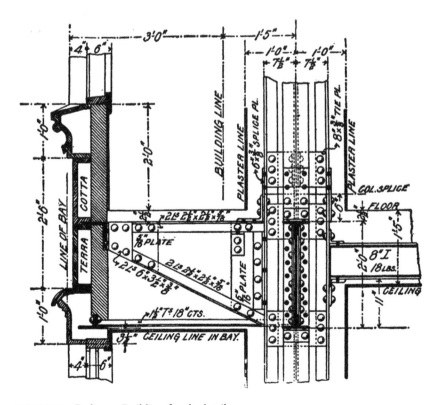

FIGURE I.3   Reliance Building facade detail.

Source: Joseph Kendall Frietag, *Architectural Engineering: With a Special Reference to High Building Construction Including Many Examples of Prominent Office Buildings*, 2nd ed., rewritten (New York: Wiley, 1912).

members or leadership who could not afford university training. The architect's relationship to the construction site that provided a setting for learning was complemented by the responsibilities of construction management. The emergence of the general contractor as an entity in the late nineteenth century absorbed the role that architects had previously provided as the coordinator of trade work and contracting. By conceding prime construction management responsibilities, architects distanced themselves from the day-to-day requirements of the construction site and, in doing so, lost skills and knowledge necessary for directing construction.

Remoteness from the construction site meant that architectural practices in the twentieth century would need to rely on different means for gaining technical knowledge. Prior to World War II, the majority of architectural staff did not enter practice with a university education, instead relying on trade schools and mentorship within practices. Individuals with academic educations received limited technical education in universities, with schools modeled on the German Polytechnic model offering more technology than those based on the French Beaux-Arts model.[19] Bound for management positions, university-trained architects relied on junior staff, including career senior staff, for technical capability. Through the first half of the twentieth century, education for non-university-educated firm members increasingly came from trade schools, although mentorship continued to play an important role.

Hierarchies within architectural practice have reinforced the notion that technology is subservient to and independent of design. This model counters a general understanding that technology, understood as hardware applied to specific tasks, is instrumental in advancing human well-being by expanding human capabilities.[20] Relegating building technology to a complementary component of design perceived more as a nuisance than an asset has undermined architects' claims of supporting humanity by undervaluing and suppressing one of its most potent dependencies. Similar to automobiles and other products such as home appliances, buildings as machines support human activity and efficacy.

Alternative definitions of technology include the notion that technology is applied knowledge and skills resonant in individuals.[21] Over the last hundred years, the history of architectural labor indicates a trajectory where technical knowledge has been deemphasized on the path toward responsibility in practice. Around the early twentieth century, the job description of architectural draftsmen came to encompass senior as well as junior architectural staff who executed technical documentation, often with pride and respect. Draftsmen, who often studied at trade schools at night, also played an important design role, especially with details. Work reductions during the Great Depression and World War II greatly depleted the ranks of experienced draftsman. After the war, a university education experience displaced trade schools as the path to the architectural office, merging technical and design education. A result was a sorting of technical and non-technical roles in practice, with a hierarchy remaining between design and production.[22]

Engineers and Technology

Technology can be defined as a sociotechnical system, such as a transportation system, in which components are linked.[23] Building can be similarly thought of as systems composed of smaller systems. This notion of technology, and corresponding difficulties in managing different building systems, has caused greater challenges for architects and architectural historians than for engineers. For example, as reinforced concrete gained traction in the early part of the twentieth century, the influence of engineers and product manufacturers on design increased, ending whatever toehold architects had left on structural design for larger buildings. Experiments of the expressive possibility of structural concrete can be found in Paris with the church of St-Jean de Montmartre and August Perret's apartments at 25 rue Franklin and garage at rue Ponthieu. Although the concrete is clad on Perret's rue Franklin apartment, it is considered significant because the facade reflects the structural potentials of the system, represented by large bays in the base and the continuation of correspondingly large infills of glazing above. The structural concrete face of the garage at the rue Ponthieu apartments is exposed, demonstrating another advantage, which is its dual role of structure and cladding.[24] Fluidity of form is an attribute initially exploited in utilitarian structures that later inspired architects, although predicting the performance of concrete is even more difficult than steel. The latter reality ensured that technical detailing of reinforced concrete was squarely in the hands of engineers.[25]

In hindsight, it is unsurprising that in a technological context increasingly dominated by engineers and industry, the communicative function of enclosure gained heightened importance and significance for architects in the postwar period. Plumbing, electrical and mechanical technologies are less compatible with objectives of formal expression than structural systems. Unlike structure, these systems are generally compact and discreetly distributed in buildings without requiring significant changes in building form and enclosures, although pipes and vents disrupted otherwise orderly facades. Prior to the war, professional mechanical engineers and air-conditioning manufacturers organized to establish and codify standards rationalized by formulas and testing, limiting solutions to those dominated by complex machines. Architects were conspicuously absent from the codifying of mechanized comfort system standards. By 1960, mechanical air conditioning became standard for commercial construction in the United States, and construction economics favored a sealed envelope.[26] Passive strategies authored by architects, such as those presented by Aladar and Victor Olgyay in their 1957 book, *Solar Control and Shading Devices*, were produced less frequently. Ironically, hermetically sealed building envelopes prompted newer insulation goals to mitigate energy costs resulting from the constant operation of mechanical equipment.

Representation, Technology and Modernism

Abstract modern architecture of the interwar period was largely one of diffuse experiments with relatively little production compared to after the war. It was after the war that the true potential of modern architecture was realized in the United States, freed from limits placed on building construction in the private sector to support manufacturing of war-related products. In Europe, and in the former colonies, modernism flourished in the wake of the Marshall Plan and the desires of newly independent countries to demonstrate vitality without mimicking the official aesthetics of the old European order. The efficacy of production and materials such as aluminum during the war effort in the United States was compatible with earlier promises of an architecture of mass production and futuristic visions of prosperity and democracy. This led to a groundswell of experimentation, in which a language of modern architecture was largely undisputed and compatible with corporate capitalism. Architects including Mies van der Rohe and Skidmore, Owings and Merrill embraced aluminum curtain walls, which became emblematic of progress through technology (Figure I.4).[27] Repetitive curtain walls were good for business, but when juxtaposed across different buildings provided little contrast, losing the impact of uniqueness.

In the United States, modern architecture included brick, stone and wood cladding more so than cementitious surfaces reminiscent of the white architecture of the twenties. Despite this material richness, modernist work of the period faced criticism that its rationality cancelled out the particulars of local culture and environment. Modernism was presented by proponents, including Philip Johnson and Henry Russell Hitchcock, as a universal response to the particulars of local conditions. This placed it against local cultural traditions and building language that supported a sense of continuity and grounding, cushioning life's hard realities. Modernism also failed to engender a sense of monumentality found in the architecture of antiquity or civic scale architecture and planning showcased at the Chicago 1892 World's Columbian Exposition. Criticism turned to the communicative function of architecture, suggesting that early modern works were banal and unnecessarily disregarded historical architectural syntax. Architects who later practiced under the postwar rubrics of Brutalism and New Brutalism sought to recover human traces on material surfaces to compensate for the loss of human touch in mass manufacturing. Honesty, which was a claim of modernism, now seemed dishonest and over produced. Alison and Peter Smithson's Brutalism relished in raw finishes and exposed building services (Figure I.5).[28]

If Brutalism and New Brutalism represented a transformation of modernism, they retained some of the same targets for criticism. For instance, traditional details absent from modernism were replaced by texture and surface irregularities,

FIGURE I.4 Lever House, Skidmore, Owings and Merrill (architects).

Source: © Ezra Stoller/Esto.

but still remained abstract and illegible to most lay individuals. This led to a sus-ceptibility of critique from those within the profession impatient with the status quo and took cues from outside architecture, including the aesthetics of Pop-art. Chief among the critics was Robert Venturi, an architect who deftly character-ized high-modern architecture as boring in his 1966 book *Complexity and Contra-diction in Architecture*.[29] His largely ironic early built work incorporated whimsical signs and symbols applied to the surface of boxy buildings (Figure I.6). Behind

FIGURE I.5 Alison and Peter Smithson, Hunstanton Secondary Modern School, 1954.

Source: RIBA Architectural Archive.

FIGURE I.6 Guild House, Venturi and Rauch (architects).

Source: Photo by author.

Venturi's facades, building technology was masked and rendered inconsequential in a reading of his buildings. Popular culture is subverted and represented to a sophisticated audience, but the work lacks a feeling of authenticity that motivated modernists.

Technology, Innovation and Reality

A loosened relationship between the material reality of building and design ideas has been traced to the Italian Renaissance and humanist Leon Battista Albertti. Following this logic, architecture design ideas are purer than those realized since they are not compromised by external forces. Notions of the gap between ideas and reality, as well as the autonomy of architecture, has fluctuated since. A contemporary corollary resonates in architecture culture in which technical and creative outcomes are divided. An example is a tendency to classify architectural assemblies and engineered systems as technological, and the balance of buildings as architecture.[30] Perceiving dichotomies as exclusive, as opposed to intertwined, is consistent with a retreat from technology that is dependent on reality and identifiable problems. An irony of modern architectural history is that technology is suppressed unless it is future oriented. This mirrors the current habit of identifying the word technology with the present and future, with older versions of technology eclipsed and not worthy of attention. Under this paradigm, technological history is largely irrelevant as progress is assigned to the steady march toward a better future aided by the latest technology.

Creating exceptional architecture does not require technical innovation, but it does require technical proficiency and effectiveness. This renders history important. Returning to Le Corbusier and the Villa Savoye, it is tempting to assign technical innovation to the project even though it is about ideas of technological representation and less about building technology. This contrasts with architects who earlier demonstrated greater interest in advancing construction and understanding the potentials of materials. For example, Frank Lloyd Wright developed a modular concrete block system while working in Southern California during the early 1920s. Rudolph Schindler, born in the same year as Le Corbusier, and apprenticed under Wright, developed a tapered version of Irving Gill's lift-slab concrete wall system (Figure I.7) for his residence in West Hollywood, designed to be structurally stable and installed with minimal labor.[31]

Balancing the value of technology with expectations of the role that it plays in determining design outcomes warrants further exploration, especially with an eye to lessons available in the past. Like the current moment, the period after World War II was one in which technology was viewed as contributing to a better life, albeit somewhat naively. This contrasts with notions of autonomy related to building design that are often extended to technology in which a better life is not a primary objective. Autonomous and deterministic technologies cause us to forget the roles individuals play in creating a better and more equitable environment.

FIGURE I.7 Wall slab installation for Rudolph Schindler's house in West Hollywood, 1922.

Source: R. M. Schindler Collection, Art, Design & Architecture Museum, University of California, Santa Barbara.

It is unlikely that technology can replace human sensibilities and communicative needs, nor is it likely that aesthetics will replace the need of sound solutions that satisfy multiple senses. As historian Carroll Pursell argues, progress is mistakenly defined as technical change that plays a deterministic role in society, sweeping us away. Ceding power to a force called technology implies the inevitable, eclipsing human agency to determine what technologies are developed and how they are adopted.[32] By questioning the notion that all current external technological influences on buildings are inevitable, and considering a broader scope of technology, architects can better appreciate available tools and lessons that have been overlooked in the rush toward the future. Such lessons can be found in postwar modernism with its sobriety and legible construction details. The authors in this book demonstrate that relatively low-tech envelopes and details of the period have been unfairly characterized as boring and unredeemable, a conclusion tainted by the modest scale of period buildings, largely caused by material rationing.

An undercurrent related to technology and architecture throughout this book is the relationship of human agency to the act of building design. In the epilogue of *Studies in Tectonic Culture*, Kenneth Frampton laments the loss of craft resulting from industrialization, and the devaluation of architecture through commodification and privatization. His stark words at the end of a volume dedicated to the expression of architecture is followed by hope in a prescription. Architects can take command of the art of building as a spatial and tectonic discipline, and they can educate their clients about alternatives to the spectacle.[33] In an era where building realization responsibilities are further stratifying and performance objectives more stringent, reengagement with lessons of the past takes on increased urgency, a notion that gives purpose to the historical lessons in *Constructing Building Enclosures*.

Book Structure and Content

The structure of the book reassesses the history of building technology through two thematic currents. The first part, *Framing Enclosures*, challenges conventional notions of architectural autotomy as a primary path to meaningful expression, arguing that postwar-era building enclosure system developments were shaped by influences on the periphery of the architectural profession, including instigators such as engineers, manufacturers and governments. Thomas Leslie begins the first section with a chapter titled "Cladding the Palazzo Lavoro: Pier Luigi Nervi and 'The Borderline Between Decoration and Structure'," where he provides insight into the development of Pier Luigi Nervi's aluminum curtain wall system for the Palazzo Lavoro in Turin. Unlike Nervi's previous works, in which concrete plays a prominent role as primary structure and enclosure, in this case a curtain wall is the primary interface with the outside. In Chapter 2, "The Decorative Modernism of Aluminum Cladding: Architecture and Industry," Tait Johnson focuses on the influence of aluminum producers on enclosure design, including close collaborations between manufactures and architects. He chronicles how modern architecture became associated with aluminum through advertising and commissions with architects, including a Detroit office building designed by Minuru Yamasaki that featured a two-story gold anodized aluminum screen outboard of the glass curtain wall. In Chapter 3, "The United Nations Secretariat: Its Glass Facades and Air Conditioning, 1947–1950," Joseph Siry provides an insightful account of the envelope of the United Nations Secretariat building, which incorporated the earliest example of large-scale mechanically powered air conditioning. This chapter investigates the different options considered for the structure's innovative air conditioning and how different key interests, including Le Corbusier, the engineers, suppliers and owners, influenced the outcome. In Chapter 4, Whitney Moon's account of the development of an experimental pneumatic demountable structure includes another governmental organization, the United States Atomic Energy Commission. In her chapter, "Victor Lundy, Walter Bird and the Promise of Pneumatic Architecture," Moon chronicles early experiments in pneumatic architecture arising from wartime programs, featuring a pneumatic mobile theater designed by Walter Lundy and engineer Walter Bird that premiered in Buenos Aires. Its literal and conceptual lightness represented the appropriation of wartime technologies. In Chapter 5, "Saarinen's Shells: The Evolving Influence of Engineering and Construction," Rob Whitehead challenges the notion of exemplar design dependent on independent genius that eclipses other key contributors. He examines the collaborative practice between Eero Saarinen & Associates and Amman & Whitney Engineers during their decade-long partnership designing concrete shell projects. A variable range of tectonic results is traced to a high level of influence from the structural engineers on Saarinen's designs. In Chapter 6, "Doing Something About the Weather: A Case for Discomfort," the default trajectory for air-conditioning integration is related to the development of concepts of comfort that help fuel and maintain standards that shape buildings in the west. Andrew Cruse contrasts the eighteenth-century bathing machine and the twentieth-century psychometric chamber developed by modern engineers to rationalize comfort standards and gain confidence and authority. Through comparing these two models, Cruse challenges the inevitability of enclosure developments after the war, establishing ground for proactivity in technical matters related to architectural form.

The second part of the book, *Assembling Constructions*, explores building projects and players through the lens of detail and function to uncover material that would typically be overlooked, in the process revealing important lessons. Subjects represent a transition from early postwar modern toward more nuanced hybrid expressions of old and new, universal and local. The resulting ambiguities in some of the subjects have caused them to go unnoticed, falling outside of historical discourse and historical survey text. In Chapter 7, "Responsive Modernism: Louis Kahn's Weiss Residence Enclosure," Clifton Fordham examines the Weiss Residence, designed by Louis Kahn and Anne Tyng in the late 1940s, before Kahn gained widespread fame. The house does not embody characteristics of the monumental in the bold manner of his later projects; rather, it represents a transition moment between universality, craft and the local. He describes how modern design can support human needs through implementing an innovative integrated window wall system that mediated light, privacy, ventilation and heat. In Chapter 8, "Prosaic Assemblies: The Rich Pragmatism of Sigurd Lewerentz and Bernt Nyberg" Matthew Hall analyzes the eccentric details of Swedish architect Sigurd Lewerentz's architecture, which defy stylistic categorization and challenge expected juxtaposition and use of materials. He describes how Lewerentz's career included an important and brief collaboration with architect Bernt Nyberg. In Chapter 9, " 'The Material of the Future': Precast Concrete at the 1962 Seattle World's Fair," Tyler Sprague shifts the focus from field-shaped material to factory-formed concrete, which permitted greater predictability and uniformity of finish. He features the United States Science

Pavilion built for the 1962 World's Fair, which represents a collaboration between architect Minoru Yamasaki and engineers John Skilling and Jack Christiansen. This chapter describes how innovative thin, prestressed concrete shapes were produced, resulting in a rib-like patterned enclosure and colonnade, anticipating the facades of the World Trade Center towers. In Chapter 10, "The Concrete Facades of Paul Rudolph's Christian Science Building, 1965–1986," Scott Murray explores the ramifications of material choices through an investigation into one of Paul Rudolph's lesser known buildings, the Christian Science Building in Champaign, Illinois, which was demolished in 1986 after only 20 years of use. The ribbed monolithic cast-in-place concrete had no added thermal insulation, which negatively impacted the energy performance and led to its abandonment and demolition. This chapter relates challenges of facade design and construction presented to the aesthetic virtues of twentieth-century modernist architecture. In Chapter 11, "Bill Hajjar's Air-Wall: A Mid-Twentieth-Century Four-Sided Double-Skin Facade," Ute Poerschke and Mahyar Hadighi investigate the work of William Hajjar, a little-known architect who developed a four-sided double-skin facade design for a seven-story office building in the early 1960s. In each story, fresh air could enter from one corner of the double-skin facade through the louvers, circulate horizontally and exhaust in the air chimney at the opposite corner. The system also included lighting fixtures and a "radiant curtain," providing a complete system for solar heating, electrical heating, passive cooling, daylighting, shading and electrical lighting. In Chapter 12, "Defining the Double-Skin Facade in the Postwar Era," Mary Ben Bonham provides a history of double-skin facades, relating them to the progression of technical, economic and resource conditions. Her journey revolves around the story of the Occidental Building, which had the first high-tech double-skin facade in the United States. In the process, she interrogates the contested definition of a double-skin facade. For Chapter 13, "Enclosure as Ecological Apparatus: Biosphere 2's 'Human Experiment'," Meredith Sattler investigates the limits of architectural expression and performance embodied in the Biosphere 2, which operated as the tightest envelope ever constructed: approximately 360 times tighter than the Space Shuttle. As a materially closed but energetically open enclosure whose performance was paramount, B2's envelope required significant engineering and architectural innovations.

Taken together, these final seven chapters provide histories that would normally fall outside the boundaries of what constitutes significant architectural history. Viewed from the perspective of enclosure and construction, they provide important pieces of a history that is key to our complex world, in which efficacy is dependent on understanding fuller notions of how to build. Building enclosures in the postwar era were economically responsible, aesthetically cohesive and comprehendible. Considering how disorienting, turbulent and fragile our current world is, the built environment, especially the parts that speak to the public, is as important as ever.

Notes

1. Jacques Sbriglio, *Le Corbusier: The Villa Savoye* (Berlin: Birkhauser, 2008), 106.
2. Rumiko Handa, *Architecture of the Incomplete, Imperfect and Impermanent: Designing and Appreciating Architecture as Nature* (New York: Routledge, 2015), 77–87.
3. Mohsen Mostafavi and David Leatherbarrow, *On Weathering: The Life of Buildings in Time* (Cambridge: MIT Press, 2003), 5–16, 82–84.
4. V. Boone and B. Gandini, "Exploring the Visual Material Within the Building Process of the Villa-Savoye," in *Building Knowledge Constructing Histories*, Vol. 1 (London: CRC Press, 2018), 373–79.
5. For a historical understanding of the representative purpose of architecture, see *Modern Architecture: Representation and Reality* by Neil Levine. For a philosophical argument on the purpose of architecture, see *The Ethical Function of Architecture* by Karsten Harries. For a practical argument, see *Why Architecture Matters* by Paul Goldberger.
6. Le Corbusier, *Towards a New Architecture* (New York: Praeger, 1974), 100–19.
7. Sbriglio, *Le Corbusier*, 98.
8. Boone and Gandini, "Exploring the Visual Material," 374–79.
9. Dankmar Alder and Louis Sullivan maintained a partnership in Chicago from 1880 to 1894 in which Adler served as an engineer, and Sullivan solely as an architect. Also, see Andrew Saint, *Architect and Engineer: A Study in Sibling Rivalry* (New Haven, CT: Yale University Press, 2007), 171–205.
10. Reyner Banham, *The Architecture of the Well-tempered Environment*, 2nd ed. (Chicago, IL: University of Chicago Press, 1984), 9.
11. Ibid., 9–15.
12. Nigel Whiteley, *Reyner Banham: Historian of the Immediate Future* (Chicago, IL: MIT Press, 2003), 109–12, 142–60.
13. Norman Tyler, Ted Ligibel, and Ilene Tyler, *Historic Preservation: An Introduction to its History, Principles, and Practice* (New York: W.W. Norton & Company, 2009), 27–42.
14. Bill Addis and Nick Bullock, "The Construction History Society," *Construction History*, Vol. 28 (The Construction History Society, 1985), i–x.
15. John Summerson, "What Is Construction History?" *Construction History* 1 (1985), 1–2.
16. Andrew Saint, "Architect and Engineer: A Study in Construction History," *Construction History* 21 (2005–6), 22.
17. Ibid., 22, 28.
18. William McDonald, "Roman Architects," in *The Architect*, ed. Spiro Kostof (New York: Oxford Press, 1977), 31, 39–44.
19. Michael J. Lewis, "The Battle Between Polytechnic and Beaux-Arts in the American University," in *Architecture School: Three Centuries of Educating Architects in North America*, ed. Joan Ockman (Cambridge: MIT Press, 2012), 67–89.
20. Stephen Kline, "What Is Technology," in *Philosophy of Technology: The Technological Condition*, eds. Robert C. Scharff and Val Dusek (Oxford: Blackwell, 2003), 210.
21. Ibid., 211.
22. For a comprehensive story on the role architectural draftsmen played in the execution and maintenance of technical knowledge, see George Johnson, *Drafting Culture: A Social History of Architectural Graphic Standards* (Cambridge: MIT Press, 2008).
23. Kline, "What Is Technology," 211.
24. William Curtis, *Modern Architecture Since 1900* (London: Phaidon Press, 1996), 76–79.
25. Amy D. Slaton, *Reinforced Concrete and the Modernization of American Building 1900–1930* (Baltimore: Johns Hopkins, 2001), 2–3, 86.
26. Gail Cooper, *Air-Conditioning America: Engineers and the Controlled Environment, 1900–1960* (Baltimore: Johns Hopkins, 1998), 167–84.

27. Gwendolyn Wright, *USA: Modern Architecture in History* (London: Reaktion Books, 2008), 156–67.
28. Alan Powers, *Britain: Modern Architecture in History* (London: Reaktion Books, 2007), 98–110.
29. Robert Venturi, *Contradiction and Complexity in Architecture*, 2nd ed. (New York: Museum of Modern Art, 1977), 16–17.
30. Banham, *The Architecture of the Well-Tempered Environment*, 9–10. Banham states that architectural historians before his time considered structure, construction materials and daylighting as part of architecture.
31. David Gebhard, *Schindler* (San Francisco: William Stout Publishers, 1997), 45.
32. Carroll Pursell, *Technology in Postwar America* (New York: Columbia University Press, 2007), iv–x.
33. Kenneth Frampton, "Postscriptum: The Tectonic Trajectory, 1903–1994," in *Studies in Tectonic Culture: The Poetics of Construction and Twentieth Century Architecture,* ed. John Cava (Cambridge: MIT Press, 1995), 382–87.

PART 1
Framing Enclosures

1

CLADDING THE *PALAZZO LAVORO*

Pier Luigi Nervi and "The Borderline Between Decoration and Structure"

Thomas Leslie

In July 1959, a committee planning celebrations for Italy's centenary in 1961 announced a competition for an exhibition structure in Turin, to house exhibits relating to the history of Italian labor. The *Palazzo Lavoro* was to be vast—45,000 square meters of column-free exhibition space. More daunting, it would have to be constructed in just ten months. Responses to the open competition drew proposals from around the world, in particular a daring shell structure by Turin architect Carlo Mollino. The organizers, however, chose a scheme by Pier Luigi Nervi that, unusually for him, proposed not a dramatic leap of concrete, but rather 16 mushroom-like structural modules: tapering piers topped by dramatically balanced cantilevering roof elements, tapering toward and separated from one another by thin strips of glass (Figure 1.1). Nervi's selection was controversial; his scheme violated the requirement that the space be spanned without columns. But he argued, successfully, that no single-span shell could be constructed in the time available. His proposal would allow construction to be staged so that each of the 16 piers could be completed on a staggered schedule. This would allow a perimeter enclosure to be assembled and built in parallel with the structure, rather than waiting until a shell was complete and had come up to strength before starting work on the enclosure.[1] Mollino was incensed, but Nervi, as both an engineer and a builder, was the only entrant who could personally guarantee completion within the time allotted.

The commission came in the midst of work leading up to the 1960 Olympics in Rome, which would put Nervi on the world stage as a designer and constructor of breathtakingly large spans imprinted with finely scaled networks of ribs. Often ignored in the glowing praise for these structures, however, was Nervi's

FIGURE 1.1 *Palazzo Lavoro*, Turin, Italy. Pier Luigi Nervi/Nervi & Bartoli, 1961. Interior view.

Source: MAXXI Museum.

mastery of daylight and the dialogue that his renowned concrete shells carried on with the glazing systems that introduced this light into his spaces. Around the perimeter of the graceful *Palazzetto dello Sport*, for instance, Nervi detailed a steel curtain wall to match the undulating edge of the arena's dome, bringing in light under the long-span roof. In a masterfully subtle detail, each bay of curtain wall gently inflects from vertical, sloping slightly outward to meet the dome's undulating edge perpendicularly.

Even more expressively, the curtain wall around the larger Olympic arena, the *Palazzo dello Sport*, featured a crisply detailed steel-and-glass skin suspended between the edges of concrete disks that shaped that building's gently arcing concourse. Punctuated by tri-lobed concrete struts that also serve as wind bracing struts on the interior, the *Palazzo's* elegant proportions made it a star; the building featured in Fellini's 1960 *La Dolce Vita*, in Michelangelo Antioni's classic 1962 film *L'Eclisse* and in a 1964 fashion shoot in *Vogue* magazine (Figure 1.2).

Nervi developed a language for these cladding systems that defined their places in his buildings' structural hierarchy. While handling fewer loads than concrete structures, cladding systems nevertheless bear the loads of their own materials and of wind. They therefore formed fertile territory for Nervi's interest in expressive structural form. Like the lifting and bracing fork-shaped buttresses that held the dome of the *Palazzetto* above its seating, elements of building cladding could, Nervi believed, be part of a grammar of static shapes and of relations between elements.

The wind braces of the *Palazzo* provide an example of Nervi's philosophy (Figure 1.3). They are deepest at the center of their spans, where the bending

FIGURE 1.2 *Palazzo dello Sport*, Rome, Italy. Pier Luigi Nervi/Nervi & Bartoli, 1960. Views of curtain wall from interior, from *Vogue*, April 1, 1964.

Source: Conde Nast.

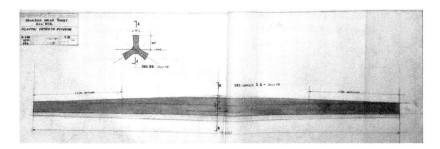

FIGURE 1.3 *Palazzo dello Sport*, Rome, Italy. Pier Luigi Nervi/Nervi & Bartoli, 1960. Shop drawing of precast curtain wall support.

Source: CSAC, University of Parma.

moment induced by wind loads on the glass outside is greatest. The detailing at their ends conveys that they are simply pinned to the concourse roof and floor. In a similar display, they taper to meet these elements, revealing that the main structure braces itself against any circumferential torsion, or radial thrusts, leaving these components elements alone to support and to resist the wind loads of the curtain wall. These are minor elements, and these detail choices are subtle, but they are important in revealing to the passerby how they operate and the hierarchy of their performance in relation to the whole. For Nervi, these details were fundamental communicative tasks and the basis of what he termed "structural architecture," the linguistic resolution of elements into systems that can be read, visually and kinesthetically, by the layperson.

The Challenge of the *Palazzo Lavoro*

In Nervi's words, the commission for the *Palazzo Lavoro* put forth "an architectural, economic, and constructive problem of unprecedented complexity."[2] His solution broke the complexity of a large span—525′ square—into 16 repetitive units, allowing for repetitive formwork, standardization of components and telescoping of construction and fabrication time.[3] Each unit consisted of a tapering concrete pier, rising from a cross-shaped plan to a circular cross section; Nervi would later explain this as a programmed structural section, offering stability at the base and flexibility at the top. To achieve these sections, Nervi relied on twisting timber formwork based on ruled surface principles— the edges of the thin timber formwork strips reified the straight lines of a mathematically conceived ruled surface, while the twisted timber itself formed a surface that negotiated between the gradually changing angles of the lines.

Nervi first used a version of this principle at the UNESCO secretariat in Paris to produce that structure's sloping, haunched piers at the ground floor. He divided the formwork for each 21 m pier into five sections, each of which could be formed by prefabricated, reusable molds of twisted timber. Each mold could be disarmed as each pier came up to strength and re-used on the next one to be built (Figure 1.4). Above these piers, Nervi was forced by time constraints to change his original proposal, of concrete roof forms. Recognizing the difficulty in forming and scaffolding such broad cantilevers, he re-conceived them in steel,

FIGURE 1.4   *Palazzo Lavoro*, Turin, Italy. Pier Luigi Nervi/Nervi & Bartoli, 1961. Construction view showing staging of pillars and steel roof.

Source: MAXXI Museum.

hiring engineer and fabricator Gino Covre to direct their calculation and production. This allowed rapid erection immediately behind the crews forming the
piers, but it also introduced a new material into Nervi's structural *oeuvre*—while
he had designed steel elements in his previous structures, nowhere had the material played such an important role in the static performance of one of his long
spans.[4]

These 16 mushroom roof elements were to be structurally independent;
glass-covered gaps between them delineated the overall structure. But, more
importantly, these separations allowed each mushroom to rotate slightly under
uneven loading or wind pressure, reducing potential bending stresses within
the roof structure. This structural freedom, however, left a serious issue at the
perimeter, where the program required full environmental enclosure. Given
the movement possible with each carefully balanced steel roof element, how
could a cladding system be supported while also allowing for such dynamic
behavior?

Gio Ponti, a multi-dimensional architect and designer, was ultimately hired
to design the interior exhibitions for the *Palazzo*. While he and Nervi did not
fully agree on the relationship of the exhibits to the punctuating structure,
they were both adamant that the pavilion had to be as transparent as possible,
to broadcast the contents of the interior to passing festival attendees and to
serve as an illuminated beacon for the exposition grounds at night.[5] Nervi
thus conceived the entire cladding system as a tautly stretched glass curtain
wall, maximizing visibility while minimizing weight. This wall would cover
two zones, separated by a gallery mezzanine level 3 m above ground. Below
this, a standard storefront system allowed for entrance doors as well as solid
and glass panels, depending on the functions behind. Above, however, this
left a 19 m span from mezzanine to roof, a height that promised significant
wind forces, and a considerable load from the weight of the glass itself. Further
complicating the skin's conception, Ponti and Nervi both recognized the need
to balance transparency with solar control—the developing exhibits featured
changes in light levels and color that risked being washed out by too much
direct sunlight.

Having expanded his palette of building materials to include structural
steel, Nervi and his architect son, Antonio, developed a curtain wall that was
as articulate in its expression of function as any of Nervi's most legible structures (Figure 1.5). The wall is formed of aluminum extrusions that hold single
thicknesses of plate glass in place with gaskets. At regular 3.75 m intervals,
these mullions are braced by concealed vertical steel plates that are, in turn,
connected by short horizontal struts to curving steel elements set 90 cm outside the plane of the skin. The mullions are shaped to reflect the bending they
absorb from wind loads, becoming deeper toward their midspans and tapering at their ends, much like the precast wind bracing struts in the *Palazzo*. To

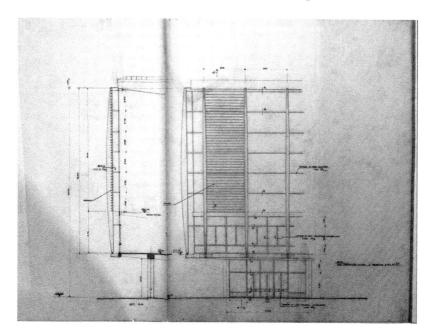

FIGURE 1.5   *Palazzo Lavoro*, Turin, Italy. Pier Luigi Nervi/Nervi & Bartoli, 1961. Design drawing of curtain wall system.

Source: CSAC, University of Parma.

achieve this shape, Nervi detailed welded steel elements, braced by internal ribs and tabs, with all external joints ground smooth to present continuous surfaces to the exterior.

At their connections—to the concrete mezzanine below, and to the steel roof above—the mullions meet articulate steel pins that transfer their horizontal loads into their respective, larger-scale structural elements while preventing transfer of bending stresses into the roof or mezzanine slabs. This strategy in itself was clever and efficient, minimizing material by shaping the wind-beams and lessening the need for reinforcement through the fuse-like pins (Figure 1.6). But Nervi saw an extra requirement here—that of visually and formally *explaining* how these elements resolved their various loads. The steel wind-beams, for instance, could easily have been integrated into the cladding plane, but Nervi saw fit to pull them out by 90 cm, providing a visual gap between them and the cladding that is bridged by the short steel struts. Furthermore, both end conditions are articulated to express this wind-bracing function.

At their bases, the wind-beams meet triangular pulpits that appear to have been gently pulled from the mezzanine slab. From below, they read as thickened regions of a perimeter girder, itself part of an articulate pattern of ribs

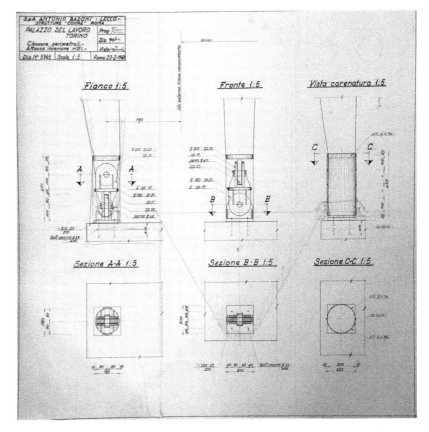

FIGURE 1.6 *Palazzo Lavoro*, Turin, Italy. Pier Luigi Nervi/Nervi & Bartoli, 1961. Design drawing of pin connections in curtain wall by fabricator Gino Covre.

Source: CSAC, University of Parma.

that expresses the isostatic patterns of stress in the loaded slab (Figure 1.7). At their tops, meeting Covre's perched steel roofs, the wind-beams are connected to articulated steel arms. The arms are pinned on both ends, restraining them laterally while letting them and the steel roofs move vertically relative to one another, due to wind forces or thermal expansion.

The two different end conditions emphasize the function of the curved steel members—they *bear* only their own weight, but they *brace* the vast expanse of glass behind them against horizontal, not vertical, forces. Crucially, they reflect the nature of the material to which they connect; the concrete slab is gently curved to accommodate the outboard connection, while the arm above is clearly a separate component.[6] On the three sides with solar exposure, Nervi

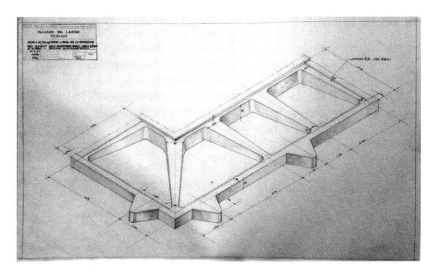

FIGURE 1.7   *Palazzo Lavoro*, Turin, Italy. Pier Luigi Nervi/Nervi & Bartoli, 1961. Design drawing of concrete frame showing triangular "pulpits" for wind-bracing system.

Source: CSAC, University of Parma.

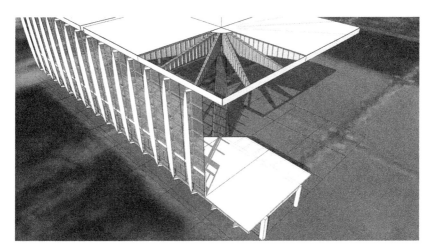

FIGURE 1.8   *Palazzo Lavoro*, Turin, Italy. Pier Luigi Nervi/Nervi & Bartoli, 1961. Digital reconstruction of curtain wall and roof assembly.

Source: Author.

detailed a system of steel louvers after carefully studying sun angles and their effects on the interior; these span from wind-beam to wind-beam, their angled position allowing them to carry their own weight across the relatively short spans (Figure 1.8).

Critical Reception

The resulting structure won global recognition for its striking scale and its nuanced expression. Its concrete pillars, for many, marked a continuation of Nervi's interest in engaging structural form; Peter Blake, writing in *Architectural Forum*, praised it as a "Concrete Parthenon," echoing the comparison of Nervi's 1960 *Palazzetto dello Sport* to the Roman Pantheon.[7] Esther McCoy, however, writing in *Arts and Architecture*, noted that with the *Palazzo*, Nervi had moved on from the "melted stone" of his earlier, "form-resistant" structures and had here taken advantage not only of steel's rapidity of construction but also its falling price in Italy. While noting that Nervi still had little patience for traditional "post-and-beam" construction, the piers and roof in particular had allowed him to match the classical order of the ancient system with his own, modern interest in letting the eye "follow the flow of forces" from one element to another, whether those elements were steel or concrete.[8] This extended to the perimeter wall, where the wind-beams, arrayed on the planning grid of the rigidly symmetrical structure inside, matched expressive static form to welded steel fabrication. The walls showed where wind forces were greatest and how these were transmitted into the roof and mezzanine diaphragms.

Ponti, meanwhile, took advantage of the skin's transparency, laying out an angular exhibition plan based on sight lines from the paths outside through the glass, framing the colossal piers inside into cinematic vistas well outside the building's perimeter. The result was a strikingly open, transparent structure that was not only a "poem in cement" but also an "ice cube out of a giant's refrigerator." *New York Times* foreign correspondent Arnaldo Cortesi, writing about Turin's postwar renaissance of building and development, highlighted the *Palazzo* in terms that emphasized the delicacy of its articulate facades: "the Turin building is a gigantic prism of glass on a square base," he wrote.

> Dimly seen through its glass walls are the Cyclopic columns supporting the roof through which shafts of light fall to the marble floor as in an ancient cathedral. . . . When the inside of the building is lighted, the beauty is enhanced. The glass sides become invisible. All that is seen is a forest of columns. [Figure 1.9][9]

Cortesi's description is revealing, as it describes one of Nervi's most subtle design strategies—the dialogue between large-scale concrete forms that do the major work of spanning large distances, and the secondary systems that enclose the spaces sheltered by these. Nervi's detailing skill allowed him to emphasize—and often to exaggerate—the vastly different proportions and characters of these two realms. Setting a thin, steel-framed curtain wall back from the edges of bulkier, more robust concrete edge members at the *Palazzetto dello Sport*, for example, makes it very clear that the enclosure has been set into the structure long after

FIGURE 1.9 *Palazzo Lavoro*, Turin, Italy. Pier Luigi Nervi/Nervi & Bartoli, 1961. Nighttime view during 1961 Fair.

Source: MAXXI Museum.

the concrete has been poured; it is secondary in scale and in function, but also temporally, in construction. At the *Palazzo Lavoro*, a handful of details emphasize the sense that the perimeter glazing is tenuously held in place, a more delicate system reliant on the larger elements of steel and concrete surrounding it for the vast majority of its support and sustenance.

Nervi was constantly interested in detail as a means of expressing the hierarchy and overall order of his structures; while the shape of individual elements formed a structural *vocabulary*, the ways in which these elements were connected to one another operated as their *punctuation*. He addressed this most explicitly in his lectures while teaching architects at the *Università di Roma* that were transcribed by Italo-American architect Roberto Einaudi while he was a student there. "There is always room for personality in the realm of technology," Nervi taught. "The artistic effect changes with a few centimeters."[10] However, within those few centimeters lay the difference between an element, component or system that was mute, and one that could explain itself to laypersons' eyes and minds in terms of its static function.

At one end of the structural spectrum—bridges, for instance, or extraordinarily long spans—the designer's personality is of necessity subordinate to the performance requirements of the structure. But at smaller scales and shorter spans, or in elements that faced minimal structural challenges, there was more room for this expression:

> If you build a vase or a chair, all forms can be used. If its dimensions are increased two times, perhaps everything will be OK. When its dimensions are increased 10 times many solutions have to be thrown away. When its dimensions are increased 100 times there remain perhaps 1 or 2 in a hundred. When its dimensions are increased 1000 times there is probably only one solution possible and perhaps none.[11]

At some level, "the borderline between decoration and structure" disappeared. For Nervi, the signal indicator for this transition was the mode of production. As soon as the methods of forming or shaping an element became important considerations, the designer's ability to manipulate form entered into the equation. If the realms of performance and fabrication had to be balanced anyway, room is left for analysis, decision, interpretation and formal expression. While explaining the rational principles behind the form of, say, the tapering piers of the *Palazzo Lavoro* was the more important aspect of that project's structural expression, the gentle curves of the curtain wall wind bracing was subject to the same demand for principle-based, meaningful and legible expression.

Seen this way, the curtain wall structure is remarkably communicative for being such a minor system in the overall scheme. The arcing shape of the braces reveals that these elements act as simply supported beams, turned upward from their usual horizontal orientation. The braces reveal that the forces they resist are lateral and distributed over the length of each brace; were they resistant to a single point load, they would properly be shaped as long, thin triangles, thickest at the point of loading. The curved shape represents a considerable investment in shaping and metalwork; surely it would have been simpler and no less effective to use rolled sections—tee or wide flanged shapes—to stay the glass and aluminum enclosure against wind. But this arrangement would have created visual uncertainty, since these shapes could be interpreted as resisting bending or simple axial loads.

Ranks of elements that could be read as columns would have suggested that, perhaps, the roof was supported by this perimeter, rather than being cantilevered from the mushroom piers within. The curved shape, however, presents far too thin a cross section at its ends to carry such loads. A fundamental difference between a column and a beam, after all, is that the axial load on a column is consistent throughout its length, while the bending load on a beam varies with distance from the support. Tapering these braces eliminates the possibility of their carrying a bearing load, leaving only the lateral bracing function as a possible interpretation.

At their bases, the attenuated pin connections further this reading: far too thin to carry loads of the magnitude being carried by the piers inside. Counterintuitively, the pins emphasize the compressive fragility of the braces, strengthening a reading of them as bending elements. If these braces themselves read as structural vocabulary, and the pins as connecting punctuation, the base and top conditions provide further articulation. The resulting material distinction between concrete can be readily formed to enable the clean assembly and layering of the systems being attached to it. Nervi's triangular pulpits provide the horizontal separation of bracing, and the curtain wall itself allows us to read the layering of louvers, support and enclosure as distinct but connected phrasing of these sub-systems into a visually coherent, functionally unitary whole.

Curtain Walls at Dartmouth and Beyond

A carefully detailed, rigorously ordered and thoughtfully expressed stanza of elements and systems formed a quiet partner to the dramatic structure and effusive exhibition design within the *Palazzo*. This relationship of elements served as the basis for further cladding systems by Nervi that borrowed these basic principles and sculpted them into unique, iterative variations on this general theme. Most closely related to the *Palazzo* were the end facades for two sport arenas at Dartmouth College: the Nathaniel Leverone Field House (1965) and the Thompson Ice Arena (1974). These structures were different in their conception from both the *Palazzo dello Sport* and the *Palazzo Lavoro*. They use extruded arches rather than domes or rectilinear prisms.

Nervi preferred linear extrusions to domes because they allowed for traveling scaffolding during construction; one section of metal scaffold could be built that spanned just a few arch modules to support formwork, while the roof was formed while it cured. The scaffold could then be rolled down the job site as sections of the roof shell were placed or poured. In a dome, the complex static performance of the doubly curved shell meant that the entire structure had to be scaffolded at once, to provide temporary support until the dome cured and could act as a monolithic, hyperstatic unit. What the extruded arch gained in constructive simplicity, however, it lost in the complexity of sealing its ends.

This was partly an architectural problem—how does a systematic method of construction based on a repetitive structural module logically conclude without obscuring the primary structural behavior within? But it also presented serious issues of deflection and creep. A long-spanning shallow arch moves considerably with loading, in particular under the asymmetrical loading caused by wind. Any cladding system connected to such an arch therefore must "ride" free of the arch's movement. Otherwise, the arch will at some point bear on the cladding itself, subjecting it to compressive forces for which it wasn't designed. Complicating matters, cladding the *end* of such an extruded shape means that the system must span a range of heights—shallow at the ends and tall in the center. Deflection varies along the arch in real terms, as do wind loads. To what extent, therefore, should the cladding's structure reflect these variations?

The *Palazzo Lavoro* system—of a planar curtain wall braced by vertical elements tuned to the type and magnitude of their carrying loads—proved to be a flexible enough idea that Nervi was later able to apply it successfully to two arched projects. At Leverone, the wind-braces take the form of triangular Vierendeel trusses, articulated by pin joints at their base and by rocker arms at their tops (Figure 1.10). These are formed from simple pipe steel, three chords set apart from one another by welded horizontal plates at regular intervals along their heights. The rocker arms allow the structural shell beyond to rise and fall with wind and thermal forces while the trusses stay in place—a flexibility that is enhanced functionally and visually by the horizontal struts that connect the trusses to the curtain wall. These are joined by pin connections at both ends, repeating the combination of horizontal fixity and vertical freedom of the top joint. They occur at the same heights across the facade, providing a regular if subtle rank of horizontal datum lines that regularize the varying dimensions of the trusses as they connect to the gentle arc of the roof. As with the *Palazzo Lavoro*'s curtain wall, the Leverone cladding visually speaks to its own purpose and fabrication *and* to its place in the structure's overall hierarchy. Its pin and rocker connections are articulated to deny any reading of compressive load. The enclosure and bracing layers are detailed to emphasize their fragility compared to the deep concrete edge to which they are attached.

Across South Park Street from the Field House, and built later, the Thompson Arena interprets this system entirely in concrete. While the extruded arch is structurally similar, Nervi (and here, more apparently, his architect son Antonio) chose to form the wind-bracing system in concrete as well. This may have reflected cost issues on the project, but the choice presented an amplification of the expressive conundrum than the steel braces on his earlier projects had—how to detail a material in two distinct ways. The expressed concrete of the *roof* had to show that it was the primary structural element, but the same material in the *curtain wall* had to visually explain that it was subsidiary, and that it in no way supported the edge of the shell above.

Ultimately, these braces were formed from precast concrete, allowing them to be thin and therefore visually delicate. But they could not approach the spidery nature of the trusses at Leverone. Instead, the detailing energy here is focused on the connections themselves. The braces are pulled away from the precast cladding system. They sit directly on an exposed ground beam, but they are connected to the cladding and the overhead spanning arch by two distinct connections. At the top of the cladding system, an expressed edge takes the shape of the arch above, connected by short nibs in the precast elements that are repeated through the height of the struts (Figure 1.11).

Above this edge, a deep shadowgap separates the cladding visually from the shell above—which is connected to the strut, again, with a steel rocker joint that is also exposed. The eye thus reads the struts and the precast cladding *both* as sitting on the edge beam below but rigidly connected to one another. They are

FIGURE 1.10   Nathaniel Leverone Field House, Dartmouth College, Hanover, NH.
Pier Luigi Nervi, 1965. Contemporary view of wind-bracing trusses.

Source: Author.

FIGURE 1.11   Thompson Ice Arena, Dartmouth College, Hanover, NH. Pier Luigi Nervi, 1975. Contemporary view of west elevation showing concrete wind bracing.

Source: Author.

both *separated*, however, from the shell above: the cladding by the deep shadow-gap, the struts by the rocker connections. Whether this consciously registers or not, the eye reads the struts and cladding as a single system. They are tenuously attached to the shell and restrained there only from falling one way or another, not in any way supported or supporting one another.

This last example is particularly subtle, and not nearly as striking as the system in its purest form—that at the *Palazzo Lavoro*. But it shows Nervi's thorough sense for articulation, expression and detail as punctuation that separated his designs into structural language. Nervi's fluency in static form will always be most apparent in the largest elements of his structures—the grand, breathtaking sweep of his arenas. But the sensibility behind the *architecture* of these structures, the translation of physical principles into experiential richness, was applicable at any scale. Indeed, Nervi's brilliance lay precisely in his ability to communicate these principles, not merely to employ them.

The most dramatic of these cladding systems was, perhaps, one that remained unbuilt. Nervi's speculative commission for an enclosed horse-racing track in Richmond, Virginia, was conceived as publicity for the Reynolds Aluminum Company. In parallel with his work on the *Palazzo Lavoro*, Nervi proposed the same basic system to enclose the open sides of an arch-shaped, folded aluminum plate shell. With a span of over 1300 feet and a height of 256 feet, the project would have been the largest enclosed space in the world. But the roof, made of lightweight, thin elements of ductile aluminum, would have been particularly lively, with an estimated deflection of up to 6 feet due to wind or thermal

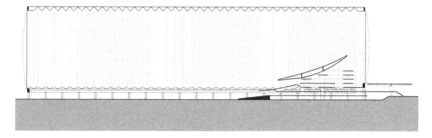

FIGURE 1.12   Project for Reynolds Aluminum. Pier Luigi Nervi, 1961. Re-constructed section showing wind bracing on end cladding.

Source: Author.

expansion. Nervi's drawings show gigantic versions of the *Palazzo*'s braces, arcing out to a depth of more than 12 feet from their narrow pin connections at the base, and massive rocker connections at their tops (Figure 1.12).

Notably, Nervi showed these braces springing not from the foundations but rather from the upper deck of a multi-tiered automotive drop-off network. This is not where they would have been most effective structurally, but where they would have been most visible. Clear of visual encumbrances from the arcing highways and viaducts below, arriving spectators would have seen through the tall glass walls to the landscape beyond. The walls braced by these vertical beams were designed to be read as one system among many in the gigantic, sublimely scaled assemblage of structure, enclosure, seating and space. The curtain wall here, like that at the *Palazzo Lavoro*, would have been dwarfed but—critically for Nervi— hardly silenced, carrying on a quieter but no less carefully considered part of the structure's dramatic conversation.

Notes

1. See Cristiana Chiorino, "Torino: Il Palazzo del Lavoro e il Ruolo della Grande Com- mittenza Industrial," in *Cantiere Nervi*, eds. Gloria Bianchino and Dario Costi (Parma; CSAC, 2012), 134–38; Massimo Ferrari, "Aule di Pier Luigi Nervi" in *Cantiere Nervi*, eds. Gloria Bianchino and Dario Costi (Parma; CSAC, 2012), 78–86.
2. Pier Luigi Nervi, "Il Palazzo del Lavoro," *L'architettura: Cronache e Storia* 7, no. 4 (August 1961): 224.
3. Roberto Einaudi, "Pier Luigi Nervi, Lecture Notes Roma 1959–60," in *La Lezione di Pier Luigi Nervi*, eds. Annalisa Trentin and Tomaso Trombetti (Milan-Turin: Pearson Italia, 2010), 112.
4. Nervi, "Il Palazzo del Lavoro," 224–37.
5. Gio Ponti, "Esposizione Internazionale del Lavoro: Impoastazione Programmatica— Documentazione e Sintesi," Typescript in Folder 36/8, Pier Luigi Nervi Archives, MAXXI, Rome.
6. Nervi, "Il Palazzo del Lavoro," 224–37.
7. Peter Blake, "Concrete Parthenon: Nervi's Prize Winning Design for the Palace of Labor at Turin," *Architectural Forum* 112 (May 1960): 112–25.

8. Esther McCoy, "A Palace of Labor—Pier Nervi [sic] and Antonio Nervi," *Arts and Architecture* (November 1960): 10–11, 32.

9. Arnaldo Cortesi, "A Construction Boom Beautifies Turin," *New York Times*, April 23, 1961.

10. Roberto Einaudi, "Pier Luigi Nervi, Lecture Notes Roma 1959–60," in *La Lezione di Pier Luigi Nervi*, eds. Annalisa Trentin and Tomaso Trombetti (Milan-Turin: Pearson Italia, 2010), 80.

11. Ibid., 146.

2

THE DECORATIVE MODERNISM OF ALUMINUM CLADDING

Architecture and Industry

Tait Johnson

Standing at the side of Highway 10 near Detroit is the former Great Lakes Regional Sales Office (1959) for Reynolds Metals, described by the company when it opened as a "Jewel on Stilts." The shimmering screen (Figure 2.1) that wraps the building was designed to attract the eye of passing motorists and executives from the automobile industry interested in the potentials of aluminum. Its architect, Minoru Yamasaki, subscribed to popular tenets aligned with modern architecture, such as a rejection of hand-carved ornament. Yamasaki did not eschew ornament completely, however, for he embraced "machine-made" ornament as a fully idiomatic expression of the age.

Yamasaki's design work, along with that of Edward Durell Stone, is considered an outlier of modernism. Categorized as *New Formalism*, it incorporates classical aesthetics, spatial strategies and proportions. Yet, his work, inclusive of ornament, is not as atypical as generally understood. Traces of ornament are more widely echoed by other practitioners throughout the twentieth century, challenging the notion that modern architecture singularly avoided ornament and decorative effects.[1] In particular, architects working with aluminum promoted decoration as the appropriate response to the landscape of consumerism spreading rapidly in the postwar United States.

The architecture firm Harrison & Abramovitz looms large in this development, working closely with aluminum producer Alcoa to promote the metal for use in architecture. Like Yamasaki, Harrison & Abramovitz was also directly hired by an aluminum producer (Alcoa) as part of a large advertisement campaign for aluminum, which included metal-clad buildings as "silent salesmen" for all things aluminum. To catch the eye, a decorative motif was not out of the question. Underpinning the embrace of decoration was a belief, shared with producers such as Alcoa and Reynolds, that aluminum possessed an inherent beauty which

FIGURE 2.1   Minoru Yamasaki, Great Lakes Regional Sales Office for Reynolds Metals, Detroit, Michigan, 1959.

Source: Photograph by Joseph Messana. University of Nebraska-Lincoln Libraries Joseph Messana Architectural Image Collection.

could be celebrated with the correct treatment. When this beauty was allowed to shine, the metal was promoted as possessing the ability to modernize the landscape and promote progress, a resonant concern growing out of the ravages of World War II. An examination of the marketing and production strategies of twentieth-century light metal manufacturers shows the ways modern architecture was spread by conscious strategies wherein architects and constructed landscapes were appropriated as domains of marketing.

Forming an Image of Aluminum

Key formative moments of the aluminum industry show how important it was to control the image of aluminum and help explain why so much effort was undertaken to promote the metal. The material properties of aluminum also contributed to a process of aesthetic rationalization. Producers framed aluminum as possessing inherent aesthetic qualities, justifying decorative treatment for designers of aluminum panels. First produced in the early 1800s, aluminum was borne of a chemical process that yielded small droplets of solidified metal. In 1825, Danish chemist Hans Christian Orsted identified the resultant yield as "a lump of metal which in color and luster somewhat resemble(s) tin."[2] The spread of aluminum products soon followed, and a sample of aluminum was displayed at the 1855 Paris Exhibition, entitled, "L'Argent de l'Argile" (the silver from clay). Early in its promotion, aluminum was associated with high value, but with a quotidian origin. Like a precious metal, it was aligned with silver, but with a humble background. This was part of an important effort to control the image of aluminum— a material that was new and widely unknown.

The process of aluminum production by chemical reaction was time consuming, costly and low in yield. A new process, utilizing electricity and backed by huge sums of capital, labor, machinery and governmental support, launched aluminum into global production at an industrial scale. The industrial manufacture of aluminum increased production scale over time and its deployment in competitive markets lowered costs, facilitating the spread of aluminum applications. An electrolytic process, developed near-simultaneously by two experimenters in 1886, formed the foundation of industrial-scale aluminum production. Paul Héroult held the patent for the process in France, and Charles Martin Hall held the patent for the process in the United States. Hall expanded production capacity and along with several investors founded a company in Pittsburgh eventually known as the Aluminum Company of America, ALCOA. By the turn of the century, industrial production of aluminum was underway in the United States and Europe.

Bauxite mining and the vast amount of electricity used for the electrolytic process made the production of aluminum mysterious and difficult to comprehend. Because aluminum is so widely useful, it is susceptible to being perceived as ersatz or valuable.[3] Aluminum can be used for staple food containers or seen as highly prized. Napoleon III, for instance, reserved utensils of aluminum, more prestigious than gold or silver, for the use of his most esteemed guests. Because of its malleable identity and relative newness, producers invested in marketing campaigns to develop and control the image of aluminum as a widely useful and important material. Recalling the association of aluminum with the precious metal silver in the 1855 Paris Exhibition, promoters identified properties of the metal that they believed led to advantages over other materials. They identified several properties of aluminum that they in turn promoted as *value*. Aluminum was understood as lightweight, durable, malleable and resistant to corrosion. Because it is lightweight and malleable, it was advertised as more easily handled by workers than brick or stone. Because it is durable and resistant to corrosion, it was promoted as a robust weather barrier to last decades. Producers also claimed aluminum possessed aesthetic value.

Marketing aluminum as beautiful was supported with its use for small-scale architectural elements, including ornamental railings and embellishments. For example, the railing and cast aluminum staircase at the Monadnock Building in Chicago (1892) is one of the earliest known uses of cast aluminum in architecture. Reacting to the work produced by the Winslow Brothers, the foundry that made the aluminum components in the building, *Ornamental Iron* declared aluminum

FIGURE 2.2 Engineers and metallurgists examine the aluminum tip of the Washington Monument in 1934, after 50 years of resisting the elements.

Source: Photograph by Theodor Horydczak. Theodor Horydczak Collection, Library of Congress, Prints & Photographs Division.

to be a "beautiful metal, which is rapidly growing in public favor."[4] While the aluminum in the Monadnock Building was overtly visible, the first use of aluminum in architecture in the United States was mostly inconspicuous. It too, however, was closely aligned with aesthetics in promotional advertising. Shortly before a 5.5 × 5.5 inch pyramidal cap of aluminum was placed at the top of the Washington Monument in 1884 (Figure 2.2), it was ceremonially displayed in a window at Tiffany's jewelry store in New York City, contextualized by a store of jewels and metals renowned for aesthetic value.[5]

Aluminum Cladding Before World War II

World War II is an important temporal marker for the spread of aluminum cladding and differentiating the aesthetics of aluminum. Before the war, aluminum was ornamental in many of the same ways as other material assemblies, formally categorized as Art Deco or the aesthetics of revival movements. After the war, decoration was most often manifest as abstracted geometric forms made reproducible by standardized manufacturing processes. Nonetheless, decorative considerations were a focus for designers before and after the war, despite the spread of theoretical strictures demanding the abandonment of ornament and decorative intent. Aluminum is a modern material, and its deployment as a decorative pattern, or ornamentation, challenges the dominant narrative about modern architecture.[6]

The earliest uses of aluminum cladding can be categorized into spandrels and panels. Spandrels were easier to install because they were a comparatively small unit that could be cast and erected on a steel and concrete frame. Several projects were underway in 1929 that vie for the first aluminum spandrel installation, such as the Koppers Building (1929) in Pittsburgh, claimed by Alcoa to be "the first large use of aluminum spandrels" with 900 units.[7]

The original buildings of Rockefeller Center (1930–39) are likewise one of the earliest large-scale uses of aluminum, with over 4400 spandrels in all (Figures 2.3 and 2.4). This project is made significant by the fact that it involved Wallace Harrison, an architect whose firm began a long working relationship with Alcoa that illustrates how producers formed mutually beneficial relationships with architects to sell aluminum. This relationship provides a narrative that begins with Rockefeller Center and continues into the postwar era with Harrison & Abramovitz's role in designing and specifying decorative aluminum cladding.

On the Rockefeller Center project, Harrison worked under the umbrella organization called Associated Architects, led by Raymond Hood. Harrison was a vector for the use of aluminum on the project. Alcoa had developed a robust sales operation and deployed their New York City salesman Fritz Close to approach Harrison. Close went to the waiting room where he hoped to catch Wallace Harrison walking in on his way to the office in the morning.[8] He subsequently sold over three million pounds of aluminum for the entire project.[9]

Along with many corners of the American social and political spheres, Alcoa planned with optimism for an enlivened postwar future.[10] Market planners with the company were aware of the lucrative possibility of expanding aluminum sales deeper into the construction market, especially emboldened by the vast production capacity available from the numerous plants constructed for the war that were operating far short of full capability immediately after the war. To help plan this postwar future, in 1945 Alcoa convened an off-site meeting of executives who recalled Wallace Harrison's role with the aluminum spandrels at Rockefeller Center. They resolved at the meeting to hire an architect with Harrison's stature and experience to help meet stringent fire codes with new cladding products so the company could sell aluminum widely across the United States in places where the fire code required a one-hour fire rating. They envisioned not only aluminum spandrels, but a whole "metal-clad building."[11]

Casting Decoration in Terms of Function

Alcoa found a precedent in a building for which Alcoa supplied aluminum for the exterior.[12] The Department of Public Works Building (1930–31), Richmond,

FIGURE 2.3   Installation of Rockefeller Center's aluminum spandrels, 1932.

Source: Photograph by Nadinejoos from Pixabay. Pixabay GmbH.

Virginia, is important for its exterior wall assembly (Figure 2.5). Alcoa executive H. F. Johnson identified the building as "probably the first American office building to use insulated metal walls just a few inches thick."[13] Alcoa featured the building in advertising to demonstrate the possibilities years before aluminum was used as cladding beyond decorative adornments or its more widespread variant, as a spandrel panel.[14] Although the thin wall was not adequate to provide the

FIGURE 2.4   Aluminum spandrels, Rockefeller Center, 1932.

Source: Photograph by Raison Descartier. Raison Descartier, licensed under Creative Commons Attribution—NoDerivs 2.0.

FIGURE 2.5   Aluminum cladding on the Department of Public Works Building, Richmond, Virginia, 1931.

Source: Photograph by Dementi Studio. Dementi Studio.

fire barrier required to meet the more stringent codes of cities like New York, one Alcoa promoter understood it as "the forerunner(s) of later prefabricated insulated aluminum panels for residential, industrial and commercial buildings."[15]

Tracing the development of insulated aluminum panels is crucial to understanding how decoration was important to the architects and designers involved in cladding because the justification for such treatment was rooted in functional characteristics of the metal and wall assembly. Similarly, a reverence for the primacy of function was never far removed from the design process for Harrison and Abramovitz. Over time, panel decorative effects starting with simple adornment blossomed into elaborate repeating patterns. Alcoa consciously cultivated their relationship with Harrison and Abramovitz and capitalized upon their creative faculties in the same way that the architects benefitted from their close association with Alcoa. Associating with well-known modern architects was a strategy that provided media coverage in the top architecture magazines and conveyed the relevance of aluminum products to decision makers in the design and construction industry.

The cladding on the Davenport Work Administration Building (1949) in Iowa was an outgrowth of a strategy to develop an exterior wall assembly that could meet stringent urban codes. Harrison and Abramovitz collaborated with Alcoa's engineers to design the building and the wall assembly, which took the form of precast, vertically ribbed aluminum panels (Figure 2.6), described by Alcoa as a "new and revolutionary type of aluminum curtain wall construction for office and institutional buildings."[16] Nine hundred and sixty-two cast aluminum panels were installed, 4×7 feet in dimension, weighing just 162 pounds each.[17] Crucially, as part of a wall assembly including sprayed perlite concrete backup, the panels were successful in passing a UL test that was adequate to meet most city building codes—the lucrative market for which Alcoa was aiming.[18]

Functional characteristics of the Davenport wall were paramount, but aesthetics went hand in hand. Reviews in *Architectural Forum* glowingly praised the image of the building, calling it a "gleaming package."[19] Alcoa described the cladding as "a highly decorative, curtain wall."[20] Davenport was labeled "decorative" but subservient to function. This hierarchical relationship served as a touchstone for the justification of decorative treatment in a line of skyscrapers that followed. Even though the Davenport building was only four stories, it was advertised as a "'pilot' skyscraper," showing that Alcoa clearly was focused on selling aluminum by the ton on large towers in the future. Harrison and Abramovitz were central figures in this ambition.

Building as Marketing

A derivative of the Davenport wall was used on the Bradford Hospital (1951), designed by Thomas K. Hendryx and selected by the hospital board "because of its modern beauty and easy maintenance." Alcoa highlighted this decision as yet

FIGURE 2.6 Ribbed aluminum cladding on the Davenport Works Administration Building, Davenport, Iowa, 1949.

Source: Photograph by Tait Johnson.

another way of equating aesthetics with function. However, the narrative that Alcoa was most keen to advertise was the role of the hospital as a pilot project for their own headquarters in Pittsburgh, the Alcoa Building (1953) by Harrison & Abramovitz (Figure 2.7). As described in Alcoa's own promotional literature, the new headquarters was designed to be "literally a thirty-story 'showcase' of aluminum construction innovations."[21] The innovation most important to reproducing an image of modernity was the exterior cladding. Photographs of the outside were reproduced in magazines and photo essays, providing further advertising and fulfilling the building's purpose as a showpiece for aluminum. Most visibly, the "diamond X" pattern of the aluminum cladding was associated with function and aesthetics. The pattern is striking, stamped on 6 × 12-foot panels that are 1/8″ thick. Named after the most popular jewel, the "diamond pattern," as it was called, was rooted in tradition but justified by function. By one account, an architect working in the office of Harrison & Abramovitz had conceived the pattern based on inspiration from a drawing he produced as a student at the Ecole des Beaux-Arts in the 1920s.[22] Max Abramovitz highlighted its origin in the function of providing panel rigidity but noted its aesthetic value, writing, "the pattern makes for a rich, textured wall, an overall visual effect which changes as

FIGURE 2.7   Harrison & Abramovitz, the Alcoa Building, Pittsburgh, 1953.

Source: Photograph by Tait Johnson.

the observer or the sun does."[23] In promotional material developed after completion, Alcoa portrayed it similarly, exclaiming, "the pyramid design was mainly specified for aesthetic reasons."[24]

Echoing Alcoa's celebration of the panel's aesthetics, Abramovitz wrote, "I feel that the joint can and will and should, in many cases, become very decorative. I don't see why it isn't as much an element of decoration and expression as the deformation of a panel." Abramovitz justified the surface effects for aesthetics and function, exclaiming, "aesthetically it can be pleasing, provide accents and contrast, and also give our structure a life of light and shade and relief that the movement of the sun is ever ready to provide."[25]

A theme of decoration continued with a subsequent commission won by Harrison & Abramovitz, promoted by Alcoa in a different light that presented decoration in subservience to function.[26] The Republic National Bank Building (1954) in Dallas, Texas (Figure 2.8), was advertised in superlative terms by Alcoa as "one of the largest and most impressive aluminum-clad skyscrapers in the nation."[27] Alcoa's marketing meshed the functional advantages of the aluminum cladding with decorative credentials, wherein "panels were impressed with a distinctive prismatic design which stiffens as well as decorates the sheet."[28] The building won praise for its panels, as noted in *Architectural Forum*, "it glitters handsomely

FIGURE 2.8 Aluminum cladding "X" pattern, Republic National Bank Building, Dallas, 1954.

Source: Photograph by Worth B. Chollar.

in the sun far across the cotton lands, and on gray days depends on its repeat pattern of embossed squares, like a fancy waistcoat."[29] Functionally, the panels were described by *Architectural Forum* as more advanced than those on the Alcoa Building, yielding what Alcoa called the "thinnest curtain wall" at 1–1/2″ including the exterior aluminum sheathing and 1–1/2″ of rigid backup insulation, advantaged by a permissive Dallas building code.[30] The Dallas system represented the achievement of Alcoa's vision—a thin curtain wall of decorative aluminum that met the building code of major cities. In 1932, Alcoa hoped that someday a 10-inch wall could be a mere 3–1/2″ made possible by aluminum.[31] Doing so would also confer an advantage of increased floor space, yielding increased rental income for the building owner.

Over time, architects specified aluminum-framed window curtain wall systems more frequently than opaque panels. Despite the decreasing cost of aluminum after the war, upfront costs remained expensive compared to other materials. However, interest in large expanses of glass from modernist architects like Mies van der Rohe helped generate opportunities that aluminum manufactures could take advantage of. Alcoa asked Harrison & Abramovitz to help develop the design of aluminum as a frame, asking the firm to "do a compatible solution with less metal" for their schematic designs for the United Nations Secretariat (1948–52).[32] Furthermore, by leveraging a loophole in building codes, glass exterior wall assemblies were not necessary to meet the same fire rating as aluminum-clad wall assemblies. Given the desire for increased floor space afforded by thin exterior walls, window offered advantages over aluminum-clad walls that were greater in dimension to achieve adequate fire ratings. Facades clad with decorative aluminum panels were also challenged by tooling costs, particularly on small projects.[33] Low cost, however, is not the sole determining factor in architecture. Styles ascend or descend in popularity and with those styles are associated forms and materials, sometimes by architects, sometimes by buyers and sometimes by the marketing prowess of material manufacturers. By the end of the 1950s, few decorative aluminum facades were planned.

Rival Producers

The spread of aluminum cladding was punctuated by World War II, an engine of change for the entire aluminum industry. This was not a solitary result of the war, as evidenced by the great quantities used on projects like Rockefeller Center, but the war was a significant cause in accelerating the use of aluminum. With government backing, Alcoa built aluminum plants all over the United States to increase the output of aluminum for the war effort, building machine gun turrets, components for munitions and airplane parts. But Alcoa wasn't the only aluminum producer engaged in the war effort. One company, Reynolds Metals, was an upstart competitor to Alcoa before, and a formidable rival after the war.

Reynolds Metals was founded by R. S. Reynolds, Sr., nephew of tobacco producer R. J. Reynolds. After initially producing tin foil wrapping for cigarette packs, the company expanded into the manufacturing of aluminum products, given greater urgency when Reynolds observed the increased level of aluminum output in prewar Germany. Reynolds argued for government support to increase the nation's wartime output capacity, which would advantageously position the company to grow in rivalry with Alcoa. Reynold's rise was buoyed by generous loans, tax advantages and the government's pursuit of Alcoa for charges of monopoly.[34] The upstart producer emerged from World War II not only having contributed significantly to the production of aluminum for the war, but also in possession of state-of-the-art plants that had been operated by Alcoa, acquired by Reynolds with government backing after the war. From this position, the company engaged concertedly in the building and construction market.

Making Aluminum Modern

To sell aluminum, Alcoa and other producers consistently advertised aluminum as aesthetically valuable. Philosopher and sociologist Georg Simmel has noted that value is not a constituent property of things but instead a subjective judgement.[35] In contrast, producers asserted that the properties of aluminum conferred value, and one of these values was an inherent beauty, worthy of decorative application. Reynolds marketed aluminum as possessing a "permanent natural beauty."[36] Likewise, Alcoa marketers wrote that aluminum cladding possessed a "modern beauty."[37]

In the postwar period, aluminum was not only sold in a modern industrial context—it was also overtly promoted as modern. For instance, Reynolds Metals commissioned a book series, *Aluminum in Modern Architecture*, which implicated aluminum as modern in association with interviews of several giants of modern architecture, including Minoru Yamasaki and Mies van der Rohe, not to mention explicitly labeling the material as "the Modern Metal."[38] Likewise, Alcoa asserted aluminum as a "modern metal" in advertisements for buildings built with the material as cladding. For the producers, a *modern* label was deeper than simple rhetoric. Instead, it was the manifestation of a deeply held conviction about the material that was rooted in their belief about the *agency* of aluminum, or, what they believed aluminum could do. What made aluminum modern, in the eyes of the producers, was its status as a superior material and their belief that aluminum could foster a better environment. To be modern was to be superior to other modes and means—superior to the old ways of fenestration, framing and enclosure, and superior to other materials. In comparison with stainless steel, an Alcoa manager said, "I think aluminum is a better material."[39]

Crucially, what underpinned the belief in the superiority of aluminum was a conviction that aluminum possessed the agency to make the world better, an ability that sprang forth from its very properties. According to Reynolds, the charge was a "never ending march" to form "the products of today and the better products of tomorrow."[40] These better products were a result of the "many natural advantages" of aluminum, claimed Reynolds.[41] These advantages were equated with abilities, and such abilities included the capability to increase rent income for buying owners because of thinner walls, claimed Alcoa, or establish specific affects and functional variations, whereby "it permits the architect to design with a more airy feeling and gives him the opportunity to vary building faces and spandrels," to quote the architect Welton Becket in *Aluminum in Modern Architecture*.[42] Alcoa advertised that the material was "of today and for the future" and "gives full expression to the modern tempo."[43] For their part, Reynolds imagined a "Brave New (aluminum) World" of greater ease, control and relaxation—welcome sentiments to a public weary from global conflict.[44] Aluminum as an escape to a better future, deployed in the present by the abilities of aluminum, was a potent narrative of the two leading producers in the United States.[45]

Delightful Billboard

Harrison & Abramovitz held a mutually beneficial relationship with Alcoa. As the firm gained commissions and developed expertise with aluminum, Alcoa leveraged their association with the firm to develop and advertise aluminum facades. Minoru Yamasaki occupied a similar position with Reynolds, designing a virtual salesman for aluminum in the Reynolds Regional Sales Office. Reynolds sought to associate with the aesthetics, ideas and figures of modernism to popularize the metal, and Yamasaki was willing to play the role. Yamasaki's openness to ornament was advantageous to Reynolds, eager to advertise what they believed was a constituent advantage and ability of aluminum—to confer aesthetic value.

Minoru Yamasaki was educated in architecture at the University of Washington and New York University. Later, he briefly worked for Harrison, Fouilhoux & Abramovitz, where he developed an admiration for the firm's modernist, design-oriented approach.[46] Today he is known for two infamous projects now demolished: the Pruitt-Igoe housing project in St. Louis (1956) and the World Trade Center in New York City (1973). In contrast to the stark and static block housing of Pruitt-Igoe, Yamasaki's modernist institutional buildings were vibrant, emphasizing narrow vertical window frames, often in harmonious dialogue with horizontal elements. Such is the case with Yamasaki's McGregor Memorial Conference Center (1958) for Wayne State University in Detroit. The McGregor Memorial committee sought a design that represented values of "beauty and repose."[47] Explaining his aesthetic aims, Yamasaki wrote that his goals were "to create a beautiful silhouette against the sky, a richness of texture and form, and a sense of peace and serenity" through spatial relationships and landscape.[48] The result was a two-story building, bisected along a center axis by an atrium flooded with natural light from skylights above. Along the exterior, aluminum-framed glass curtain walls lie behind a colonnade of marble-clad columns. Decorative aluminum screens are affixed between columns and at the ends of the atrium, providing an interplay of light and transparency.

The Reynolds Regional Sales Office outside Detroit shares an emphasis on light, geometry and materiality. At three stories instead of two, it is more commanding in elevation than the Wayne State building, made even more striking by a gold, anodized aluminum screen (Figure 2.9) that wraps the upper two stories.[49] This screen is composed of aluminum rings, 10″ in diameter, 2″ deep. Functionally, the screen blocks direct sunlight on the aluminum-framed glass curtain wall behind and extends out over a colonnade around the perimeter, providing protection from the elements for a walkway and the glass walls at the ground floor. Aesthetically, the screen was designed as a purposeful pattern of rings, two layers deep, each offset from the other to create an interlocking, decorative effect. Yamasaki's screen was called an "architecture of delight," a screen producing a play of light and shadow over the glass and aluminum facade behind.[50] Recalling its label, a "jewel on stilts," here the jewel was allowable, but only subservient to the glass box. In alignment with Harrison & Abramovitz, the other leading architects of aluminum, Yamasaki had no reservations about decorative treatment, but only if there were some justification of function and that it be machine made. Defining a marketing narrative, Reynolds declared the effects of Yamasaki's jewel "both beauty and function" and promoted the entire project as modern.[51]

Conclusion

Concerning the relationship of architecture to media, Beatriz Colomina has written, "to think about modern architecture must be to pass back and forth between the question of space and the question of representation."[52] This perspective is a useful framework for understanding the interaction between architects and aluminum producers. The producers were continually engaged in the process of representation. Most often, they advertised the potentials of aluminum in architecture through print media. But they also successfully appropriated the talent and fame of architects, while labeling the whole project—from the identity of the architects to the identity of the metal itself—modern. Furthermore, the role of the leading aluminum producers shows the great lengths to which they engaged in defining a narrative about the limits and ambitions of modern architecture. For the producers, modernism was a marketing project. The postwar United States, emerging from the horrors of war and economic depression, was a fertile ground to spread the message, resonant with visions of future prosperity and a commercial landscape of beautiful aluminum.

Contemporary commentators on architecture identify a return to ornament as an ascendant architectural trend. At the turn of the twenty-first century, Brent Brolin highlighted the "banishment & return" of ornament in architecture.[53] More recent publications, such as *The Function of Ornament*, identify the affects of manufacturing techniques using CNC routers and lasers that allow designers

FIGURE 2.9   Detail of the anodized aluminum screen surrounding the Great Lakes Regional Sales Office, Detroit, Michigan, 1959.

Source: Photograph by Joseph Messana. University of Nebraska-Lincoln Libraries Joseph Messana Architectural Image Collection.

greater decorative expression on exterior surfaces.[54] Yet, for twentieth-century aluminum cladding manufacturers and the architects who worked closely with them, decoration has long been an acceptable expression for architecture. The reasons for this are etched in economics, pragmatics and beliefs about the agency of materiality. Together, these reasons underpinned the spread of decorative metal cladding during the mid-twentieth century—a trend that lasted until such deployments became economically unsustainable in the United States.

Notes

1. Scholars have argued for a wider categorical scope of modern architecture. See Sarah Williams Goldhagen, "Something to Talk About: Modernism, Discourse, Style," *Journal of the Society of Architectural Historians* 64 (2005): 149.
2. George David Smith, *From Monopoly to Competition: The Transformation of Alcoa 1888–1986* (Cambridge: Cambridge University Press, 1988), 4.
3. For a discussion of the malleable identity of aluminum, see Robert Friedel, "The Psychology of Aluminum," working paper, 1975, Department of Science, Johns Hopkins University, Baltimore, MD., folder 2, box 51, Records of the Aluminum Company of America; Eric Schatzberg, "Symbolic Culture and Technological Change: The Cultural History of Aluminum as an Industrial Material," *Enterprise & Society* 4, no. 2 (2003).
4. *Ornamental Iron* 1, no. 5 (Chicago: The Winslow Brothers Publishers, October 1893): 87.
5. "Rival to Older Metals," *New York Times*, November 25, 1884, 8.
6. Understanding aluminum as modern relies on its entanglement in the capitalist, industrial economy. Concerning this economic context, Gwendolyn Wright maintains that many formal expressions were deployed to give form to modernism. See Gwendolyn Wright, *The Politics of Design in French Colonial Urbanism* (Chicago, IL: University of Chicago Press, 1991), 10.
7. *Let's Look at the Record* (Pittsburgh: William G. Johnston Company, 1944), folder 6, box 129, Records of the Aluminum Company of America.
8. Fritz Close, February 2, 1982, cited in Victoria Newhouse, *Wallace K. Harrison, Architect* (New York: Rizzoli, 1989), 146–47.
9. Smith, *From Monopoly to Competition,* 337. The use of aluminum spandrels was checked off by John D. Rockefeller Jr. after the managers for the Rockefeller Center, Todd & Brown, employed "professors and metallurgists" to "carefully examine existing aluminum installations to render a report on the advisability of using aluminum." See C. E. Magill, Aluminum Company of America, to Zantzinger, Borie & Medary, Architects, July 9, 1932, p. 5, folder 4, box 104, Records of the Aluminum Company of America.
10. For a description of the architectural context of postwar planning, see Andrew M. Shanken, *194X: Architecture, Planning, and Consumer Culture on the American Homefront* (Minneapolis: University of Minnesota Press, 2009).
11. *Summary of the Minutes of the Architectural Sales Meeting* (Pittsburgh: Aluminum Company of America, March 12–14, 1945), 12–13, folder 1, box 117, Records of the Aluminum Company of America.
12. Abramovitz noted that precedent could also be found in the iron and steel-skinned factories and "the more recent use of metal panels in shop fronts." Max Abramovitz, "Of UN, Alcoa Bldg. and Davenport," (lecture, Detroit Chapter of the American Institute of Architects, February 9, 1954) p. 4, folder 23, box 6, Max Abramovitz,

Architectural Records and Papers Collection (New York: Department of Drawings and Archives, Avery Architectural and Fine Arts Library, Columbia University).

13. Eugene Russell Myers, "The Development of Mid-20th-Century American Metal-and-Glass Architecture in the Curtain Wall Style" (PhD diss., University of Pittsburgh, 1963), 80.

14. Alcoa's advertisement of this building was included in a promotional brochure. See *Let's Look at the Record*.

15. Myers, "The Development of Mid-20th-Century American Metal-and-Glass Architecture in the Curtain Wall Style," 80.

16. Public Relations Department, *1949 Developments in Aluminum* (Pittsburgh: Aluminum Company of America, 1949), 1, box 50, Records of the Aluminum Company of America.

17. Abramovitz, "Of UN, Alcoa Bldg. and Davenport," 6.

18. Alcoa wrote that this wall assembly, including sprayed perlite concrete backup, attained more than twice the requirements of Pittsburgh's notably stringent building code. *Aluminum on the Skyline* (Pittsburgh: Aluminum Company of America, 1953), 4.

19. "Aluminum: New ALCOA Administration Building at the Davenport Plant is a Gleaming Package of the Many Mature Uses of this Metal in Building," *Architectural Forum* (June 1949): 2.

20. Public Relations Department, *1949 Developments in Aluminum,* 1.

21. *Aluminum on the Skyline,* 3.

22. Oscar Nitzchke interview, quoted in Newhouse, *Wallace K. Harrison, Architect,* 146–47.

23. Description of significant firm projects, p. 10, folder 3, box 1, Abramovitz Architectural Records and Papers Collection.

24. *Aluminum on the Skyline,* 9.

25. *Metal Curtain Walls* (Washington, DC: Building Research Institute, Division of Engineering and Industrial Research, National Academy of Sciences, National Research Council, 1955), 62.

26. The project was completed in collaboration with Gill & Harrell, Architects.

27. *Architectural Achievements: Republic National Bank Building* (Pittsburgh: Aluminum Company of America, 1954), folder 19, box 126, Records of the Aluminum Company of America.

28. Ibid.

29. Professor Thrugg interview, quoted in "Buildings in Review: Schizophrenic Building," *Architectural Forum* (February 1955): 126.

30. *Architectural Achievements: Republic National Bank Building.*

31. Myers, "The Development of Mid-20th-Century American Metal-and-Glass Architecture in the Curtain Wall Style," 81.

32. Abramovitz, "Of UN, Alcoa Bldg. and Davenport," 7.

33. Ibid. Max Abramovitz asserted that tooling cost for producing the jigs that stamp the decorative patterns is "infinitesimal" compared to the entire cost of a large project, but this implied that the same tooling costs would be less affordable on small projects.

34. The postwar United States was an expanding market economy presciently described by Lizabeth Cohen as a "Consumer's Republic." Cohen argues that mass consumption arose in the postwar United States via "complex shared commitment(s)" in which the interactions between policymakers, business and labor leaders and civic groups gave rise to the consumer economy. See Lizabeth Cohen, *A Consumers' Republic: The Politics of Mass Consumption in Postwar America* (New York: Vintage Books, 2004), 11.

35. Georg Simmel, *The Philosophy of Money* (London: Routledge & Kegan Paul, 1978), 73.

36. *Reynolds Aluminum and the People Who Make It* (Richmond: Reynolds Aluminum Company, 1970), 2, Reynolds Metals Company Collection, series 7, Virginia Historical Society, Richmond.

37. *Architectural Achievements: Bradford Hospital* (Pittsburgh: Aluminum Company of America, 1954), folder 3, box 126, Records of the Aluminum Company of America.

38. John Peter, *Aluminum in Modern Architecture,* Vol. I (New York and Louisville: distributed by Reinhold Pub. Corp., 1956), 9.

39. *Metal Curtain Walls,* 163.

40. "Aluminum on the March," accessed January 13, 2013, https://youtu.be/KJRD2Q7 i624?list=WL&t=1644.

41. *A-B-C's of Aluminum,* (Louisville: Reynolds Metals, 1950), 27.

42. Peter, *Aluminum in Modern Architecture,* Vol. I, 240.

43. Alcoa Advertisement, *Architecture Forum* (May 1936), 76, quoted in Myers, "The Development of Mid-20th-Century American Metal-and-Glass Architecture in the Curtain Wall Style," 160.

44. "Brave New (Aluminum) World," *Reynolds Review* (January–February 1961): 10.

45. Aluminum was marketed as an escape to the future, an idea resonant with the recent trauma of World War II. Postwar architecture as a response to the "trauma of war" is explored in Beatriz Colomina, *Domesticity at War* (Cambridge: MIT Press, 2007), 56.

46. Dale Allen Gyure, *Minoru Yamasaki: Humanist Architecture for a Modernist World* (New Haven, CT and London: Yale University Press, 2017), 8.

47. Ibid., 73.

48. Minoru Yamasaki, *A Life in Architecture* (New York: Weatherhill, 1979), 43.

49. For further discussion about the analogy of wrapping, see Grace Ong Yan, "Wrapping Aluminum at the Reynolds Metals Company," *Design and Culture* 4, no. 3 (2012): 299–323.

50. John Peter, ed., *Aluminum in Modern Architecture '60* (Richmond: Reynolds Metals Company, 1960), 73.

51. Reynolds Metals Company, 1959 Annual Report, March 1, 1960, back matter, folder 60, Reynolds Metals Company Collection, series 1.2.

52. Beatriz Colomina, *Privacy and Publicity: Modern Architecture as Mass Media* (Cambridge: MIT Press, 1994), 13–14.

53. Brent C. Brolin, *Architectural Ornament: Banishment and Return* (New York: W.W. Norton & Company, 2000).

54. Farshid Moussavi and Michael Kubo, eds., *The Function of Ornament* (Barcelona: Actar, 2006).

3

THE UNITED NATIONS SECRETARIAT

Its Glass Facades and Air Conditioning, 1947–1950

Joseph M. Siry

Defining images of mid-century American architecture feature the steel-and-glass modernism of Ludwig Mies van der Rohe; Skidmore, Owings and Merrill; Pietro Belluschi; and others. Their aesthetic of expansive glass walls forced collaborating engineers to develop approaches to air conditioning that were visually, spatially and functionally compatible. Such glass towers were unlike the interwar tall buildings, which were still largely enclosed in masonry and had punched window openings. This chapter revisits the mid-century modernist tall building typology from the perspective of how mechanical systems enabled the new style by creating habitable interiors behind glass walls. Accomplishing this feat of controlled environments within a context of nascent technologies, and little precedent, was challenging. Although it has not been emphasized in most histories, mid-century modernist architects collaborated closely with mechanical as well as structural engineers. The visual precision of a steel-and-glass vocabulary, both inside and out, forced architects to think even more closely about the role of air conditioning as an essential part of a new definition of architectural enclosure that was both material and functional.

In the canonical postwar modernist buildings, one prominent and highly visible innovation was the curtain wall. A curtain wall, as the term has been conventionally used since the 1950s, is an enclosure of glass suspended as a continuous surface outside the structural frame.[1] Additional characteristics include thin profiled window frames, the repetitive use of the glass modules and a lack of masonry between glass sections. In the mid-twentieth century, the realization of a modern glass curtain wall first appeared in Wallace Harrison and Max Abramovitz's United Nations Secretariat, for which design began early in 1947, and whose construction finished in 1950 (Figure 3.1). Upon its completion, and even before, the UN Secretariat was at the center of debates about modernism

FIGURE 3.1 Wallace Harrison and Max Abramovitz (architects); Syska and Hennessy (mechanical engineers), United Nations Secretariat, on the East River, 42nd to 48th Streets, New York City, 1947–50.

Source: Photograph by J. Alex Langley, published in *Architectural Forum* 93, no. 5 (November 1950): 102.

generally, and in particular about larger glass-walled buildings that depended on air conditioning for habitability.

Planning the Secretariat

Formed in the aftermath of World War II as a peace-keeping organization, the United Nations decided to maintain its headquarters in the United States following an invitation from Congress late in 1945. After a competition between cities, the UN's location was settled when, in December 1946, the General Assembly ratified the choice of a Manhattan site along the East River from 42nd to 48th Streets, which had been purchased for the UN by John D. Rockefeller, Jr., earlier that month. The total UN campus was New York City's largest building project after Rockefeller Center, to which it was regularly compared.[2] As Victoria Newhouse and George Dudley have described, for the design of the UN's buildings, it was assumed that architects from different member countries would participate, with an American heading the team. On January 2, 1947, the UN's first Secretary-General, Norwegian lawyer Trygve Lie, appointed Harrison as the director of planning for the United Nations' Permanent Headquarters.[3] Lie and Harrison soon appointed a Board of Design of ten architects from member countries, including most famously Le Corbusier of France and Oscar Niemeyer of Brazil. These architects met daily from February 17 through early June 1947, and soon decided that the United Nations Headquarters should contain three main buildings: the General Assembly, the Secretariat and a Conference Building for councils and committees.

By May, a preliminary consensus on the site plan and building massing had emerged around a scheme mainly by Le Corbusier, although modified by Niemeyer. Within this site plan, Wallace Harrison led in the development of a definitive design, "Scheme 53," for the Secretariat, which would tower above the General Assembly Building and the adjacent low rectangular conference building on the United Nations' 17-acre campus. While some of the earlier schemes had included multiple towers, the singular built Secretariat was a symbol of the organization's aspiration to world unity.[4] While there had been debate about whether the new buildings should be traditional or modern in style, the modernist impulse aligned with the desire to create an image of an institution that would represent the global postwar human future. Harrison said in 1947:

> For the people who have lived through Dunkerque, Warsaw, Stalingrad and Iwo Jima, may we build so simply, honestly and cleanly that it will inspire the United Nations, who are today building a new world, to build this world on the same pattern.[5]

Harrison and Abramovitz designed a 39-story rectangular slab above street level with three stories below. It is 544 feet high, 287 feet long and 73 feet wide, with 5,200 heat-absorbing operable, double-hung windows above 5,200 glass spandrels fronting low opaque walls below the windows. The windows light 840,000 square feet of space.[6] To ensure a maximal amount of sun and natural light, the Design Board decided that the building's main east and west elevations would have glass curtain walls. The narrow north and south end walls would be faced with gray Vermont marble. From the start it was assumed that the Secretariat would be air conditioned. The building's high ratio of surface area to interior volume would mean that most of the heat gain would be from solar radiation. Yet at the time the Secretariat's design took shape in 1947, before it was presented to the General Assembly in September, there was no completed tall office building with a glass curtain wall like the Secretariat. To assess heat gain, Harrison initiated several studies by his collaborating mechanical engineers, Syska and Hennessy, to determine the number of hours of sunlight the building might expect each year and what measures could be taken to minimize unwanted solar effects. The Secretariat was given a more north–south orientation, in order to minimize the shadow it would cast on the UN campus site, conceived as a sunlit expanse of verdure.[7] This orientation also avoided blocking the narrow site from 42nd Street on its south.[8] Aligned with Manhattan's grid, the Secretariat's orientation was 29 degrees east of north, so that its west wall would face northwest and thus not be as directly exposed to the sun as a wall facing due west (Figures 3.2 and 3.1). Still, this wall was of greatest concern, because in New York City, summer cooling loads peak at about 4:00 p.m.

Another alternative was to align the Secretariat along the east–west 42nd Street (Figure 3.2), which would have meant solar gain mainly through a southwest-facing wall, in addition to having the tall building cast a shadow over the campus to the north. In the late afternoon, such a southwest wall would have gained 144 Btu per hour per square foot of unshaded glass. As built, the broad northwest wall gained 122 Btu per hour. Thus, engineers designed for the corresponding maximum cooling load of 2,300 tons predicted to occur in late July (the "installed tonnage" shown by curve number 1 on the graph at right in Figure 3.2). This corresponded to the orientation as built, with long northwest and southeast walls. When the Secretariat opened, "the planners believe they have installed enough capacity to insure summer comfort for all workers."[9] Air conditioning and heating cost about $3 million in 1950 dollars, or $6.00 per square foot, compared to $4.50 to $5.00 per square foot for local office buildings, which averaged 22 percent of their facades for light openings, or about a third of the Secretariat's 68 percent.

For the testing of the Secretariat's exterior glass walls, Harrison again turned to Syska and Hennessy. In their office, Edward J. Benesch was in charge of the project. Their heat tests, with an experimental setup of single-pane plate glass versus double-layered heat-resistant Thermopane, oriented as they would be in the Secretariat, showed that the Thermopane resulted in interior temperatures 10 to 15 degrees lower than those behind the plate glass window. Thus, though 50 percent more expensive, Thermopane, with its distinctive blue-green color, was at

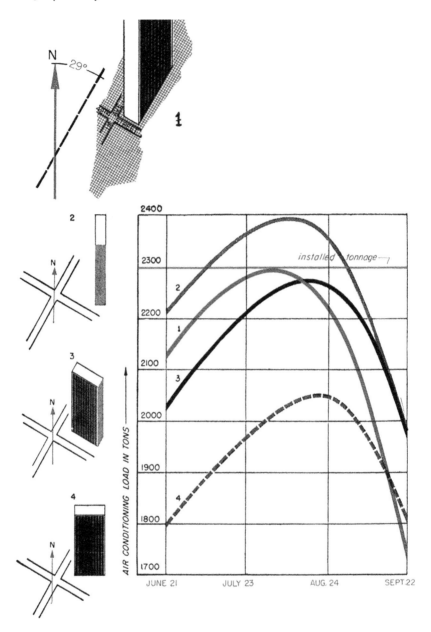

FIGURE 3.2   Harrison and Abramovitz (architects); Syska and Hennessy (mechanical engineers), United Nations Secretariat, alternative orientations and their respective peak air-conditioning loads through the summer months; (1) as built, 29 degrees east of north; (2) due north–south; (3) long wall facing southwest along 42nd Street; (4) long wall facing due south. Study by Syska and Hennessy for the UN Headquarters Planning Office.

Source: *Architectural Forum* 93, no. 5 (November 1950): 108.

first chosen to moderate heat and cold on the Secretariat's west side, with clear plate glass imagined for the east front. For aesthetic unity, it was later decided to use the more expensive tinted, double-layered, insulated Thermopane on both facades.[10] Yet Thermopane was eliminated from the specifications, due in part to its cost premium over Solex single-thickness plate glass, which absorbed heat. The Solex glass' green tint blocked infrared rays and reduced the internal temperature by 10°F relative to clear glass when the sun shined.[11] Also, windows on both sides were not sealed but were rather operable sliding sash, and the layered glass was too heavy for double-hung sashes. The windows, framed in non-oxidizing exterior aluminum mullions backed by steel, enclosed the enormous east and west walls. The engineers wrote in 1947: "Shading or screening out of the solar radiation or sun entering the conditioned space has great bearing on the capacity of the air-conditioning system."[12] Yet the only solar protection would be dark gray interior venetian blinds for all the offices. These decisions about the windows shaped the air-conditioning system, and they also had major consequences for the building's environmental performance.

The Mechanical System

The Secretariat's system was among the largest early applications of the Carrier Conduit Weathermaster System, or air-conditioning units under individual windows, designed for multi-room buildings, such as hotels, apartments and hospitals, as well as office buildings. The system was first brought out in the spring of 1941 before Pearl Harbor and used in only eight buildings during World War II, including the Statler Hotel in Washington (1941) and the Pentagon (1942).[13] Such buildings had a high percentage of outside rooms, with shifting solar heat gain representing a high proportion of the total cooling load. A system that required less duct space and enabled individual room control was ideal. To circulate 100 percent of the air required for cooling, earlier systems in office buildings had all air cooling and heating at large air handling units. Then the air, after being conditioned by the units, had been distributed vertically between floors and then horizontally through large ducts on each floor to individual offices.

With the Weathermaster system, the building's primary air-cooling was also done at a central refrigeration plant. But instead of that plant processing 100 percent of the air circulated in the building, the plant preconditioned only that 25 percent of the circulated air needed for ventilation. The system propelled this "primary air" at high velocities through narrow tubular ducts to the room units, as shown in a diagram of a window unit in the slightly later Lever House in New York City, completed in 1952 (Figure 3.3). These 6.5-inch-diameter tubular ducts were no larger than an ordinary steam pipe, or one seventh to one ninth the cross-sectional area of low-velocity air ducts, thus saving space.[14] This air entered the unit's base, where it was drawn up over finned coils of hot or chilled water. This primary air's high velocity through nozzles served as the motive power to induce secondary room air to flow over the unit's water coils to heat or cool the 75 percent or so of recirculated air. In extreme weather, one room might require twice the amount of heating or cooling required by another, and during fall and spring, one room might need cooling while another needed heating. In northern climates, intermediate weather, where both heating and cooling were required, amounted to about one-third of the days of the entire year. Different needs of occupants in mild weather had been addressed by opening windows. But that had meant noise, dust and uncontrolled gusts of outside air. The Weathermaster was designed to replace the open window for interior climate control in mild weather.

In the UN Secretariat, on each floor, double-hung windows rose eight feet above the sill. The spandrel below the sill and above the floor slab was a dwarf wall of cinder block 2 feet and 9 inches high, much like the Weathermaster unit construction at the slightly later Lever House (Figures 3.3 and 3.4). This was

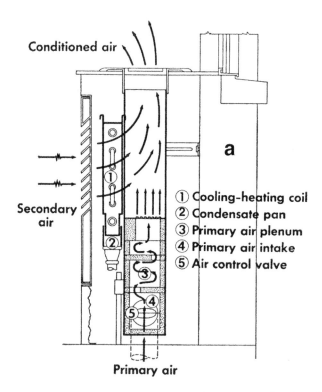

FIGURE 3.3 Skidmore, Owings and Merrill (architects); Gordon Bunshaft (chief designer); Jaros, Baum and Bolles (mechanical engineers), Lever House, Park Avenue, 53rd to 54th Streets, New York, 1950–52, diagram of a section through a Carrier Conduit Weathermaster window unit, with the outdoors at right and indoors at left, showing (a) location of dwarf wall of cinder block faced with wire-glass.

Source: © ASHRAE, www.ashrae.org, *Refrigerating Engineering* 61, no. 4 (April 1953): 389.

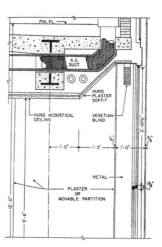

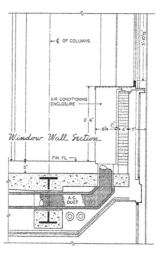

FIGURE 3.4 Harrison and Abramovitz (architects); Syska and Hennessy (mechanical engineers), United Nations Secretariat, (*left*) interior of typical office and (*right*) wall sections showing air-conditioning duct supplying high-velocity air to the Weathermaster units inside window sills.

Source: From *Progressive Architecture* 31, no. 2 (February 1950): 66. Photograph by Gottscho-Schleisner; drawings not attributed.

required by New York City's building code in a glass wall to increase resistance to fire spreading up from floor to floor. This opaque dwarf wall backed the Weathermaster unit below the sill. Externally, this wall was painted black, with a 3-inch air space between it and darker fixed heat-absorbing wire glass above the floor slab, distinct in color and translucence from the blue-green glass windows above (Figure 3.1).[15] The room's occupant set the desired temperature with a dial on the unit, which automatically maintained that temperature by controlling the flow of hot or cold water into the pipes in the unit into which room air was inducted. Independent local control yielded a savings of 15 to 25 percent in fuel costs over a system with no individual controls.

In multi-room buildings, larger air handlers occupied much rentable space, the value of which often equaled 20 to 30 percent of the cost of the air-conditioning installation. In a 15-story building, larger air handlers might occupy an area equal to an entire floor. The Weathermaster system's small cabinet unit in an individual room occupied less space than an ordinary radiator. In a rental building, the system would result in a gain of about 15 percent in rentable space, or in a 40-story building like the UN Secretariat, the equal of about two extra floors. The Weathermaster was much quieter than earlier systems, with no moving parts in the unit, so it was acceptable in private offices, hospitals and hotel rooms, where low noise levels are most needed. Installation was also much easier because the system's piping, fittings and connections were all standardized and factory-made, so there was no sheet metal duct fabrication on site. This process had been time consuming and hard to estimate, and air ducts crafted together on site were prone to leak. Air and water conduits were much smaller than big ducts, so they could be set inside hollow pilasters in buildings. As the room unit replaced the large air-handling unit, so the conduit replaced the large duct. The system was expensive to install but economical in its operating costs and space needs.

A Matter of Comfort

The Secretariat had 4,000 Weathermaster units along the outer windows, supplied with high-velocity air to cool peripheral offices within 12 to 16 feet of the windows, while low-velocity air was supplied through larger ceiling ducts for the central office areas (Figure 3.5). Syska and Hennessy presented at least five schemes for placing the units along the exterior walls.[16] After many experiments with different numbers of units, it was decided to place six Weathermaster units along the outer walls in each 28-foot-long seven-window structural bay between the columns (Figure 3.6). At every other unit were hand-operated controls. One editor noted, "such a luxurious standard, with individual controls at every second window unit, is enforced on the UN by the contiguity of Icelanders and Abyssinians in the same building, each with his own idea of thermal comfort."[17] Many individual controls gave the design unusual flexibility, because the bays were used for offices of different sizes that could change from year to year.[18] The system aimed to make conditions "climatically perfect," so that "United Nations employees, who come from many different climates, will be able to walk to any of the individual controls and by a simple adjustment, regulate the temperature to their liking."[19] To accommodate personnel with varied ideas of comfort, the temperature range for private offices was wider than usual, with the window

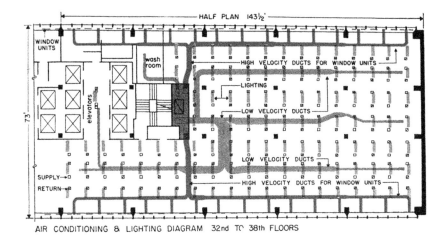

AIR CONDITIONING & LIGHTING DIAGRAM 32nd TO 38th FLOORS

FIGURE 3.5 Harrison and Abramovitz (architects); Syska and Hennessy (mechanical engineers), United Nations Secretariat, half plan showing high-velocity ducts for window units and low-velocity ducts for interior offices, with lighting fixtures and air diffusers.

Source: From *Architectural Forum* 93, no. 5 (November 1950): 109.

FIGURE 3.6 Harrison and Abramovitz (architects); Syska and Hennessy (mechanical engineers), United Nations Secretariat, piping details for the Weathermaster window units.

Source: © ASHRAE, www.ashrae.org, *Heating and Ventilating* 46, no. 12 (December 1949): 60.

units providing a spread of over 12°F.[20] The building's core also had individual thermostats that enabled control of temperature and humidity. Though it is hard to fathom today, one aim was to provide sufficient air changes to make smoking possible for all chambers.[21]

Carrier's experts promised that "occupants of adjoining offices will be able . . . to maintain any temperature they choose. What would happen if, say, an Ecuadorean and an Icelander shared one cubicle, the experts did not say."[22] This was a concern for delegates who were then still meeting in the UN's interim headquarters in the former Sperry Gyroscope Company's factory at Lake Success, Long Island. For months after taking up work there in August 1946, most expressed dissatisfaction with the "blasting air-conditioning" and the "windowless offices."[23] They complained about it in meetings where diplomats from New Zealand, India, Iraq, Iran and Nicaragua had sat side by side. A New Zealand delegate "once threatened to wear mittens to meetings unless something was done about the icy blasts, which he called 'violent Antarctic blizzards.'" A French representative "customarily arrived at the sessions clad in a sweater and white blanket clearly marked 'U. S. Medical Corps.'"[24] These conditions remained a bother during the organization's five-year residency at Lake Success, when the UN was beginning to establish its procedures amid an array of postwar international tensions. There was concern that the organization would not prove viable, and air conditioning became a flash point for dissatisfactions that had multiple causes. After a vociferous session of the Political and Security Committee in May 1947, Trygve Lie ordered that all references to complaints about the "blizzard" conditions created by air conditioning be cut out of the verbatim report. It was the first time that the text of an important UN public meeting had been amended.[25]

The leadership promised that something would be done, and Lie surely recalled such concerns when working with Harrison and Abramovitz on the new Secretariat. As the first of the new UN buildings to rise on the East River site, the Secretariat would have "the largest air-conditioning system ever equipped with individual controls." The president of the Milwaukee company overseeing the equipment's installation asserted: "Office temperatures . . . must never be a 'barrier to United Nations harmony.'" The manual controls would enable virtually every occupant to select their own weather indoors. There was naturally concern about the high initial cost of such a system, since, unlike a commercial landlord, the UN could not recover those costs from paying tenants over time. So the high initial cost would have to be justified in terms of lower-than-normal operating costs. It was stressed that in addition to the convenience of individual controls, installations of the Weathermaster system in other buildings through 1949 "have been demonstrated to cut heating costs as much as 25 percent or more."[26]

Refrigeration and air-conditioning equipment in the Secretariat's third basement supplied chilled water for the entire UN campus and conditioned air for the first basement (which contained offices) and the ground floor (Figure 3.7). The second basement was devoted to shipping and receiving, while the first

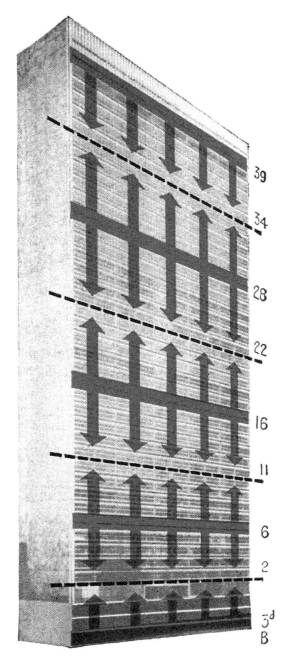

39
34
28
22
16
11
6
2
3^d
B

FIGURE 3.7 Harrison and Abramovitz (architects); Syska and Hennessy (mechanical engineers), United Nations Secretariat, diagram showing intermediate floors for distribution of air conditioning.

Source: Photograph by J. Alex Langley (as in Figure 3.1). From *Architectural Forum* 93, no. 5 (November 1950): 108.

basement was for document production and other services. Above, three inter-mediate floors (the sixth, sixteenth and twenty-eighth) had air-handling equip-ment that distributed filtered, cooled and dehumidified air to intervening floors above and below. The locations of the mechanical floors were influenced by the need to limit the size of chilled water risers and air-conditioning ducts. Each mechanical floor also held equipment, like elevator machinery, that was not part of the air conditioning. When the Secretariat opened, about 26 percent of its net area was used as mechanical and service space.[27] On average, each system sup-plied ten floors, or five above and five below. Thus, the mechanical equipment room on the sixth floor served the second through tenth floors; the eleventh to twenty-first floors were supplied from the sixteenth floor; and the twenty-second to thirty-third floors from the twenty-eighth floor. The pipe gallery in the top-most thirty-ninth floor supplied the floors just below (Figure 3.7), including the penthouse apartment of the UN Secretary General. Another in the third base-ment below ground served the entrance lobbies and council chambers. Project-ing vertical louvers in the glazing pattern mark these mechanical floors along the east and west fronts, in addition to a screen for rooftop equipment (Figure 3.1).[28] Thus, on the outside, as little as possible was done to disrupt the image of the UN Secretariat as a prismatic tower, in accord with Le Corbusier's preference for pure geometric forms. Benesch wrote, "the unique exterior architectural pattern of the building at the 6th, 16th, 28th, and 39th floors, disguises the various fresh air intakes and exhaust air discharges."[29]

The Weathermaster near-window units were arranged so that for each group of floors supplied from a pipe gallery, there were two systems for the east side and two for the west. This enabled differential heating and cooling for the northeast, northwest, southeast and southwest peripheries, whose solar heat gain would vary most through the day. For interior zones, less exposed to solar gain, one air-conditioning system served the north half and the other system the south half (Figure 3.5). The engineers wrote, "the heat gains in this area will be much less than those for the exterior area consequently the air velocity in the ducts will not be as high."[30] Each system had an air-handling unit with a dehumidifier, damp-ers, pre-heaters and re-heaters, air filters and distributing ductwork and diffusers. Chilled water was supplied to each dehumidifier from the refrigeration plant in the basement, which had two centrifugal compressors supplying 2,300 tons of cooling.[31] Cool condenser water was drawn from the East River and returned to it after use, rather than recirculating it through a cooling tower, thus saving cost but not energy.[32]

The Secretariat's Weathermaster system exemplified the trend of the 1950s, when engineers nationally were gradually changing their design criteria in favor of increasing use of high-velocity air distribution. In earlier office buildings, high-velocity air reduced shafts and furring spaces and cut down on floor-to-floor heights without changing ceiling height. For tall office buildings, high-velocity air effectively provided greater flexibility of design and reduced first costs of building. This quality became ever more valued as the cost per pound of air-conditioning ducts and the cost per cubic foot of building space approximately tripled between 1930 and 1956. High-velocity systems were also quieter and more easily adapted to individual room control, as with Weathermaster units. The use of larger windows, plate glass walls and the increased intensity of electrical illumination all increased cooling loads, which in turn demanded larger, higher-capacity air-conditioning systems. In office buildings, rentable area and usable spatial volume were at a premium, to insure a maximal return on investment. As the cost of space to enclose mechanical components rose, the virtues of high-velocity air proved more attractive, even if it was more demanding in energy consumption and thus operating expenses.[33]

Performance in Context

For the UN Secretariat's air conditioning, there was a strong organizational rationale and a carefully wrought architectural solution. But even the most careful efforts for the building itself could not control its atmospheric context along the East River. This was then still mainly an industrial district whose pollutants determined the quality of the air taken into Harrison and Abramovitz's building's system, which pulled in smoke and noxious gases that commercial office buildings on Park Avenue or Broadway did not. As the Secretariat was being occupied in 1950, Trygve Lie voiced objections to the particle and fume-laden air that was wafting toward the building's air intakes from the waterside electrical generating plant of the Consolidated Edison Company, between 34th and 41st Streets, nine hundred feet to the south of the UN site (Figure 3.8). The plant burned coal to make steam for turbines that generated electricity, and its smokestack was 75 feet lower than the Secretariat, so emissions drifted directly up and toward the UN building's air intakes. In June 1950, as the Secretariat rose, Lie had complained to New York's mayor, William O'Dwyer, that the plant's gas and smoke nuisance was so bad as to create a "serious and potentially dangerous condition."[34] The UN encouraged the utility to replace coal with natural gas as a smokeless fuel. Yet clean-burning gas would produce a preponderance of water vapor that might cause some difficulties with the air conditioning, because its dehumidification load would be increased.[35] The Secretariat's air-conditioning apparatus included air washers, but the commission thought that these would not protect the building from discoloration inside, and that emissions would stain and blacken the pristine exterior. Most critically, Lie foresaw that the air-conditioning system would carry smoke and gases into the principal council chambers.[36] City officials were concerned that the emissions would damage the marble cladding on the end walls. The power company responded that the UN should consider the possibility of waterproofing or other protection for the marble facing.[37]

Acting Mayor Vincent Impellitteri referred the problem to Robert Moses, the City Construction Coordinator, who was to confer with the UN and

FIGURE 3.8   Harrison and Abramovitz (architects); Syska and Hennessy (mechanical engineers), United Nations Secretariat, view from the north, August 25, 1952, showing Consolidated Edison Co. Power Plant 900 feet to the south.

Source: United Nations Photo Library, Photo Reference Number: JG, 57011.

Consolidated Edison.[38] The UN leaders saw the situation in part as an opportunity to demonstrate their ability to mediate disputes and reach an amicable understanding with Consolidated Edison. At a meeting where Moses presided, the UN agreed to relocate the intakes of its air-conditioning system in the new Secretariat below the thirty-ninth story, down to a level where danger from sulfur dioxide fumes was considered negligible. It also agreed to study the idea of installing new air filters to trap impurities from the power plant. The power company consented to investigate the possibility of burning pulverized anthracite coal instead of soft bituminous coal, which emitted large quantities of sulfur dioxide. It also agreed to take more sulfur dioxide concentration readings on the thirty-ninth floor.[39] These and other negotiated compromises saved the Secretariat architecturally and environmentally. Its surfaces and services had needed political support to survive.[40]

The Secretariat's great glass walls and the building's dependence on large quantities of air conditioning were criticized almost from the moment the plans were unveiled (Figure 3.9). On November 19, Le Corbusier wrote to Warren Austin, a senator from Vermont and Chairman of the Headquarters Advisory Committee, recounting his efforts to realize the glass wall equipped with *brise-soleil*, or fixed

FIGURE 3.9   Harrison and Abramovitz (architects); Syska and Hennessy (mechanical engineers), United Nations Secretariat, detailed view of base and entrance from southwest, showing sixth-floor mechanical level.

Source: © Ezra Stoller/Esto.

exterior sun shading which he had long advocated and experimented with in his own building designs. Of the Secretariat's curtain wall, Le Corbusier wrote: "I affirm that it appears to me senseless to build in New York, whose climate is terrible in summer, glass wall sections that are not equipped with brise-soleil."[41] The design was to maintain a maximal summer indoor temperature of 78°F, at peak outdoor conditions and a relative humidity of 50 percent the year round.[42] But there were condensation problems and concerns about worker comfort and air-conditioning costs. In May 1964, a four-hour power failure highlighted the Secretariat's extreme mechanical dependence. With outside temperatures nearing 90°F, the staff experienced disruptions to their work. Lights, certain elevators, typewriters and copying machines were out of service, as were refrigeration equipment and air conditioning, including that serving the Secretary General U. Thant's offices on the thirty-eighth floor.[43]

The Curtain Wall After Occupancy

Soon after occupancy, there were problems due to heat gain and glare through the east and west facades. These were most acute on the Secretariat's east side, facing the East River, where there is also considerable reflected light off the expanse of the water, and where no buildings shaded the Secretariat. The wall suffered from air and moisture infiltration, condensation that led to energy loss, occupant discomfort and visible deterioration of its structural elements, including the steel behind the aluminum window mullions. In 2010, the UN campus underwent its first full renovation, including its outmoded and inefficient mechanical systems.[44] By 1980, to reduce air-conditioning costs in summer and heating costs in winter, solar film had been installed on the east wall's windows, except for the crowning thirty-eighth floor. But the internal darkening effect and external mirror-like appearance drew criticism and discouraged the film's installation on the building's west side facing Manhattan.[45] The curtain wall was wholly rebuilt, with all glass fixed and designed to match the spectral characteristics of the original. The new walls have insulated double-thickness panels of laminated glass that include a low-emissivity coating to reduce solar heat gain, and a blue-green tinted substrate on outer glass panels. An overhaul of the mechanical systems included daylight dimming of artificial lights, demand-control ventilation, a computerized building management system and a revamping of the central refrigeration and heating plant for the whole UN campus. These changes would cut its energy use by at least 50 percent and reduce heating and cooling energy use by 65 percent.[46] Every effort was made to ensure the restored glass curtain wall's viability indefinitely. Thus, the Secretariat's materials and mechanical systems, which represented the most advanced thinking of the mid-twentieth century, have been restored, both to preserve the building's iconic image and to meet twenty-first-century standards of sustainability.

Notes

1. Robert L. Davison, "Curtain Walls," *Architectural Forum* 92, no. 3 (March 1950): 81–96; William Dudley Hunt, Jr., *The Contemporary Curtain Wall* (New York: F.W. Dodge Corporation, 1958); David Yoemans, "The Pre-History of the Curtain Wall," *Construction History* 14 (1998): 59–82; and Thomas Leslie, Saranya Panchaseelan, Shawn Barron and Paolo Orlando, "Deep Space, Thin Walls: Environmental and Material Precursors to the Postwar Skyscraper," *Journal of Society of Architectural Historians* 77, no. 1 (March 2018): 77–96.

   On the curtain wall and office culture, see Reinhold Martin, *The Organizational Complex: Architecture, Media, and Corporate Space* (Cambridge, MA: MIT Press, 2003), 95–105.
2. "Work on U. N. Site Behind Schedule," *New York Times* [hereafter *NYT*], November 25, 1948, 3. Charlene Mires, *Capital of the World: The Race to Host the United Nations* (New York: New York University Press, 2013) and Pamela Hanlon, *A Worldly Affair: New York, the United Nations, and the Story Behind Their Unlikely Bond* (New York: Fordham University Press, Empire State Editions, 2017).
3. Victoria Newhouse, *Wallace Harrison, Architect* (New York: Rizzoli, 1989), 112–13; and George A. Dudley, *A Workshop for Peace: Designing the United Nations Headquarters* (New York and Cambridge, MA: Architectural History Foundation and the MIT Press, 1994).
4. Its shadow-casting, and those of nearby buildings on the UN campus, was studied with a heliodon, a device for showing the sun's apparent motion. Gertrude Samuels, "What Kind of Capitol for the U. N.?" *NYT* Magazine, April 20, 1947, 3. Le Corbusier designed smooth glass curtain walls in his unbuilt project for the Central Union of Consumer Cooperatives or Centrosoyuz (Tsentrosoyuz) Building in Moscow (1928–36) and the built Cité de Refuge or hostel for the Salvation Army in Paris (1929–33). See Rosa Urbano Gutiérrez, " 'Pierre, revoir tout le système fenêtres': Le Corbusier and the Development of Glazing and Air-Conditioning Technology with the Mur Neutralisant (1928–1933)," *Construction History* 27 (2012): 107–28; and Brian Brace Taylor, *Le Corbusier, the City of Refuge, Paris, 1929–1933* (Chicago, IL: University of Chicago Press, 1987), 111–22.
5. Harrison, quoted in "What Kind of Capitol for the U. N.?" 59. See also "Plea for U. N. Home Moves Delegates," *NYT*, September 24, 1947, 4; "U. N. Headquarters, Progress Report," *Progressive Architecture* 31, no. 6 (June 1950): 59. The design was published in *Report to the General Assembly of the United Nations by the Secretary-General on the Permanent Headquarters of the United Nations* (Lake Success, NY: United Nations, June 1947).
6. E[dward]. J. Benesch, "Heating, Ventilating and Air Conditioning the Secretariat of the United Nations," *Heating and Ventilating* 46, no. 12 (December 1949): 57–62; and "The Secretariat, A Campanile, A Cliff of Glass, A Great Debate," *Architectural Forum* 93, no. 5 (November 1950): 94–112.
7. "U. N. Headquarters, Progress Report," 61. On the curtain walls, see Alexandra Quantrill, "The Aesthetics of Precision: Environmental Management and Technique in the Architecture of Enclosure, 1946–1986" (PhD diss., Columbia University, 2017), Chap. 1, 26–77.
8. "United Nations Builds a Vast Marble Frame for Two Enormous Windows," *Architectural Forum* 90, no. 6 (June 1949): 83.
9. "The Secretariat: A Campanile, A Cliff of Glass, A Great Debate," *Architectural Forum* 93, no. 5 (November 1950): 109. One ton of refrigeration is the amount of energy needed to freeze one ton of water at 32°F into one ton of ice at 32°F in a 24-hour period.
10. Newhouse, *Wallace Harrison*, 128.

11. Harrison, talk to Royal Institute of British Architects, London, February 20, 1951, 12. Box 4, Folder 5: U.N. (United Nations Building). Series II: Collection II (1989.003), Subseries I: Professional Papers. Wallace Harrison Papers, Avery Architectural and Fine Arts Library, Columbia University.

12. John F. Hennessy, "Preliminary Report on Mechanical and Electrical Equipment; Air Conditioning," in *Engineering and Technical Studies of the Headquarters Planning Office* (Lake Success, NY: United Nations, August 1947), 52.

13. Willis H. Carrier, "Cooling by Conduit Saves Space," *Ice and Refrigeration* 101 (July 1941): 73–76; Walter A. Grant, "From '36 to '56: Air Conditioning Comes of Age," *Heating, Piping, and Air Conditioning* 29, no. 3 (March 1957): 155–60. On the Pentagon, see Terry Mitchell, "Air Conditioning System for the New War Department Building," *Refrigerating Engineering* 42, no. 2 (August 1941): 88–89. See also *The Carrier Conduit Weathermaster System in the Hotel Statler, Washington* (Syracuse, NY: Carrier Corporation, 1950).

14. "Piped Air Conditioning," *Business Week*, June 14, 1941, 68; "Production; Air Conditioning," *Business Week*, June 21, 1941, 46; "Carrier's Conduit," *Business Week*, December 9, 1944, 76; and "Air Conditioning Spreads to Skyscrapers," *Business Week*, July 23, 1949, 26.

15. Reyner Banham, *The Architecture of the Well-Tempered Environment*, 2nd ed. (Chicago, IL: University of Chicago Press, 1984), 224–26. On the Secretariat, see "Skyscrapers Here Get Air-Conditioning," *NYT*, April 15, 1949, 42. On the windows, see also Lou R. Crandall, "Builders View Problems of Constructing United Nations Headquarters," *Civil Engineering* 20, no. 2 (February 1950): 90; and "The United Nations Secretariat," *Engineer* 191, no. 4961 (February 23, 1951): 265–66. In the 2010 restoration, dwarf walls were eliminated, and new insulated glass spandrels have painted aluminum backing to recall the walls' opaque surfaces. Nadine M. Post, "United Nations Renovation Complicated by Working Within Operational Campus," *Engineering News-Record* 270, no. 8 (March 25, 2013): 30–36.

16. United Nations, Headquarters Planning Office, Special Meeting on Secretariat Air Conditioning, June 29, 1948. Box S—0542–0045: Non-Registry Files of the Director—17. Air Conditioning—April 1948 to February 1950, Folder 24380. United Nations Archives, New York. Syska Hennessy's consultation is documented in AG-025 United Nations Registry Section (1946–1979), Headquarters Planning, Box S-0472-0030—S-0472-0031.

17. "United Nations Builds a Vast Marble Frame," 85. See also "World Capital to Have 'Push-Button Weather,'" *NYT*, August 10, 1947, 20.

18. "United Nations Secretariat," 110. Each floor's gross area was about 19,000 sq. ft.

19. Kathleen Teltsch, "Climate à la Carte in U. N.'s New Home," *NYT*, September 5, 1949, 8.

20. "United Nations Secretariat," 110.

21. Ibid.

22. Ibid.

23. "Plea for U. N. Home Moves Delegates," 4.

24. Teltsch, "Climate à la Carte in U. N.'s New Home," 8.

25. George Barrett, "Diplomat Boils Up in U. N.'s Icy Wind," *NYT*, May 11, 1947, 28.

26. Teltsch, "Climate à la Carte in U. N.'s New Home," 8.

27. "United Nations Secretariat," 110.

28. Banham, *Well-Tempered Environment*, 221–25.

29. Benesch, "Heating, Ventilating and Air Conditioning the Secretariat of the United Nations," 58. Le Corbusier had foreseen such a solution, advising that pipe gallery levels could be top levels of smaller buildings and intermediate levels of the higher ones, "so you can get a level unity through the site—levels of horizontality." Le Corbusier, Board of Design, Meeting 14, Monday, March 10, 1947, in Dudley, *Workshop for Peace*, 113.

30. Hennessy, "Preliminary Report on Mechanical and Electrical Equipment," 55.
31. Benesch, "Heating, Ventilating and Air Conditioning the Secretariat of the United Nations," 59.
32. "U. N. Headquarters, Progress Report," 59. See also "World Capital to Have 'Push-Button Weather'," 20; Edward Passmore, "The New Headquarters of the United Nations," *Journal of the Royal Institute of British Architects* 57 (July 1950): 347; and Alfred R. Zipser, "Buffalo Forge Company Leads in Industrial Air Conditioning," *NYT*, February 13, 1956, 34.
33. C. Milton Wilson, "High Velocity Air Conditioning: Its Effect on Building Design," *Architectural Record* 119, no. 5 (May 1956): 227–31.
34. "'Air Wash' Sought for U. N. Building," *NYT*, September 19, 1950, 33. See also "U. N. Capital Plans Stress Function," *NYT*, May 22, 1947, 19; "Utility Sidesteps U. N. Plea on Smoke," *NYT*, September 24, 1950, 18; and "City Seeks to Ease Smoke Evil at U. N.," *NYT*, August 8, 1950, 22.
35. "'Air Wash' Sought for U. N. Building," 33.
36. "Truman Rejects City's Plea on Gas," *NYT*, August 5, 1950, 17.
37. "Utility Sidesteps U. N. Plea on Smoke," 18.
38. "'Air Wash' Sought for U. N. Building," 33.
39. "U. N. and Public Utility Find Way to Peace; Clear Up Smoke Problem, and Cheaply, Too," *NYT*, November 22, 1950, 23.
40. In 2004, the Consolidated Edison site was sold for development, and the old power plant was to be demolished. "A Place Apart Becomes a Place Discovered," *NYT*, June 19, 2005, J11; "Dipping City's Toes into the East River," *NYT*, November 10, 2005, E1; and Jeff Vandam, "A Lot to Soak Up, Even Outside the Bars," *NYT*, April 19, 2009, RE7.
41. "Une Lettre de Le Corbusier à propos du gratte-ciel de l'O. N. U.," *L'Architecture d'aujourd'hui* 2, no. 3 (December 1950): IX. See also Le Corbusier to Warren Austin, n.d., quoted in "United Nations Secretariat," 108; and Le Corbusier, *UN Headquarters* (New York: Reinhold, 1947). The original correspondence is in Folders 31–38, Box 15, Series V United Nations, Max Abramovitz Architectural Records and Papers, 1925–1990, Avery Architectural and Fine Arts Library, Columbia University, New York.
42. Benesch, "Heating, Ventilating and Air Conditioning the Secretariat of the United Nations," 58.
43. "Power Failure at U. N.; World Body Perspires," *NYT*, May 9, 1964, 15.
44. Neil Macfarquhar, "Renovating the U. N., With Hints of Green," *NYT*, November 22, 2008, C1; Joann Gonchar, "Revival of an Icon: The United Nations Renovation Team Brings Back the Long-Faded Luster of the Secretariat While Satisfying Ambitious Performance Goals," *Architectural Record* 200, no. 9 (September 2012): 106–12; Post, "United Nations Renovation"; and *The United Nations at 70: Restoration and Renewal: The Seventieth Anniversary of the United Nations and the Restoration of the New York Headquarters* (New York: Rizzoli, 2015).
45. Brian Urquhart to Wallace Harrison, August 14, 1980. Box 6, Folders 5: United Nations—Art Committee & Miscellaneous. Series II: Collection II (1989.003), Subseries I: Professional Papers. Wallace Harrison Papers, Avery Architectural and Fine Arts Library, Columbia University.
46. Post, "United Nations Renovation," 30–36.

4

VICTOR LUNDY, WALTER BIRD AND THE PROMISE OF PNEUMATIC ARCHITECTURE

Whitney Moon

At the 1st International Colloquium on Pneumatic Structures in Stuttgart, Germany, in 1967, architect Victor Lundy presented two recently completed pneumatic projects: the traveling Atomic Energy Commission (AEC) Pavilion (1960) and the Brass Rail Pavilions for the New York World's Fair (1964–65). Carried out in partnership with engineer Walter Bird, who was also speaking at the conference, Lundy stated that pneumatics "will revolutionize the construction of the future."[1] He added, "In this sense, pneumatic structures offer great possibilities in the hands of creative architects and engineers to make new forms of wonder and possibility that are truly of our time."[2] What Lundy stressed in Stuttgart was not only the sculptural potential of this novel enclosure system but also the importance for architects to guide its future development into something more than "just great bubbles or Quonset huts over swimming pools."[3]

These two collaborations between Lundy and Bird index not only the rise and acceptance of ephemeral enclosures as alternatives to permanent buildings—specifically, they illustrate the promise of pneumatic architecture in the mid-twentieth century. That is, both the AEC and Brass Rail Pavilions demonstrated how air structures could offer both an economical and formally inventive alternative to conventional construction materials and methods. Lundy, who had not previously worked with pneumatics, understood their untapped potential for the design of temporary and mobile structures. As he explained in Stuttgart, "I see air supported structures as a great new vehicle of making giant architectural sculpture and spaces for man—with big, brilliant strokes. It is like a new palette for an artist."[4] Strategically, Lundy sought out Bird's pneumatic expertise to carry out these two different commissions.

Bird, who founded Birdair Structures, Inc., in 1956, was a pioneer in the development and popularization of air structures. Well known for his radomes—pneumatic radar domes (Figure 4.1) comprised of a thin, non-metallic protective covering designed and constructed for the US Air Force in the mid-1940s—Bird demonstrated that these unique enclosures were "capable of satisfying service requirements which can be met by no other type of structure."[5] In the next few years, over one hundred of Bird's air-supported radomes were constructed in the extreme climatic conditions of the northern frontier (i.e., severe ice, cold and wind loads), demonstrating through their successful performance the credibility and practicality of pneumatic structures.[6]

But Bird's contributions to their research and development extended far beyond military application. In 1956, the engineer formed Birdair, Inc., and initiated the first commercial applications of pneumatic structures: inflatable warehouses, sports facility covers and pool enclosures.[7] Soon thereafter, Bird's home in Buffalo, NY—or, more specifically, his pool, surrounded by an all-season air-supported enclosure—was featured on the front cover of LIFE magazine.[8] In demonstrating the capacity for inflatables to serve a wide range of practical uses,

FIGURE 4.1   Walter Bird, photographed standing on top of one of his first pneumatic "radome" prototypes on the Cornell Aeronautical Laboratory grounds in Buffalo, New York, 1948.

Source: Birdair, Inc.

Bird suggested their architectural potential as an alternative form of building envelope:

> The air structure is a unique type of structure. To take full advantage of its tremendous potential, architects and engineers must be willing to work with a new set of parameters. There are free-flowing lines, not boxes; open spaces, not individual cubicles.[9]

Despite the lack of a "universal agreement on what to call these new buildings," Bird defines pneumatic structures "as any of the wide variety of structures using pressurized air to stiffen or stabilize a flexible material to form a structural shape."[10] Distinguishing between "air structures" (single walled enclosures whose air pressure is kept slight above ambient) and "air-inflated structures" (tubular or cellular constructed walls or roofs whose air pressure lends structural stiffness), Bird explains that the popularity of his pool enclosure allowed Birdair, Inc., to develop a range of commercial, as well as more "unusual" or specialized, applications for pneumatic structures.[11]

Although it could be said that Bird played an instrumental role in both the development and popularization of air structures, they still remained a novelty throughout the 1960s. The lack of building codes and standards to regulate their construction meant that each one was handled as a unique condition, raising concerns about their safety and viability as buildings.[12] Despite the challenges that architects and engineers faced in seeing their pneumatic visions come to life, the respective expertise afforded by these two complimentary disciplines was essential in terms of advancing air structures, both technically and formally.

Because pneumatics presented a number of technical challenges, both engineers and architects typically approached them with minimal formal innovation. Yet, as Lundy explained in Stuttgart, "I believe in controlling and guiding the air-supported structure, not in standing by and witnessing its inevitability."[13] What Lundy was referring to was the propensity for pneumatic structures to resist becoming anything other than "big, round bubbular shapes."[14] Motivated by the sculptural possibilities of architecture, Lundy viewed the air-inflated membrane as an opportunity for invention.[15] This chapter investigates an alternative historiography of pneumatic architecture—one that is focused on the groundbreaking collaborations between an engineer (Bird) and an architect (Lundy), who, by working with air, together rewrote the future of constructing building enclosures.

AEC Pavilion

As part of the US Atoms for Peace program, the pneumatic Atomic Energy Commission (AEC) Pavilion was a demountable pneumatic structure designed by Lundy and Bird (Figure 4.2).[16] It traveled around South America and the world for nearly ten years to promote nuclear research and to educate the general

FIGURE 4.2 Time-lapse photographs by Victor Lundy illustrating the 1960 deflation of the *AEC Pavilion* in Rio de Janeiro, Brazil.

Source: Courtesy of Victor A. Lundy Archive, Library of Congress, Prints & Photographs Division, LC-DIG-ds-11760.

public about peacetime uses for nuclear energy. Through a series of informative, educational and interactive displays, it demonstrated the atom at work in medicine, industry, agriculture and power. After its premier in Buenos Aires, Argentina, in November 1960, the AEC Pavilion traveled to: Rio de Janeiro, Brazil; Lima, Peru; Santiago, Chile; and Montevideo, Uruguay. In April 1962, it was erected in Mexico City, Mexico, then "reconditioned" by Birdair Structures before going on an extended two-year tour to: Dublin, Ireland; Ankara, Turkey; Tehran, Iran; Baghdad, Iraq; and Tunis, Tunisia.[17]

Housed inside Lundy and Bird's 22,000 square foot inflatable enclosure and described as "a combination of scientific laboratory, training institute and display, featuring operating nuclear devices, models and lectures, and staffed by leading physicists, radio-biologists, chemists, engineers and technicians," the AEC's *Atoms at Work* exhibition was designed to communicate to the general public how atomic energy contributes to "the betterment of mankind."[18] As it toured from one location to the next, the exhibition also allowed scientists in various host countries to directly engage with a nuclear reactor and other related equipment, with support from American staff. *Atoms at Work* also offered opportunities for students and teachers to learn more about the latest developments in modern and nuclear science.

Accommodating scientists, researchers, students, and the general public, the pavilion's interior featured two primary exhibition spaces: (1) a technical center, comprised of a working nuclear reactor and laboratory located inside an insulated inflatable dome, as well as an adjacent gamma irradiation pool; and (2) a public area, including a three-screen film theater, lecture demonstration areas and informative displays (Figure 4.3).[19] Covering a wide range of peaceful uses of atomic energy, the exhibits also included several nuclear devices, including an isotope-producing training and research reactor, as well as a pool-type gamma irradiation facility.[20] As explained by Albert H. Woods, who was in charge of overseeing the entire AEC Pavilion project, including the selection of Lundy as its architect:

> After they had seen the film, the audience could move on to the technical area, and watch the guest scientists at work behind a separating barrier, with the feeling of a "glimpse behind the scenes." There would also be a series of lecture-demonstrations on the peaceful uses of nuclear energy.[21]

Inside the pavilion, a nuclear reactor was housed in another custom inflatable dome, also manufactured by Birdair. The intent of this "set" was to provide an interactive space for scientists and researchers to conduct experiments and to explore the peacetime uses of nuclear energy. As Lundy explains:

> A special, transparent, plastic bubble—an air structure within an air structure—encloses the reactor, extending almost to the top of the dome.

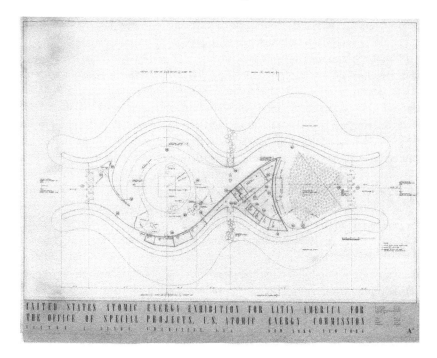

FIGURE 4.3   Victor A. Lundy, air-supported exhibition building for the US Atomic
Energy Commission, floor plan, 1960.

Source: Courtesy of Victor A. Lundy Archive, Library of Congress, Prints & Photographs Division,
LC-DIG-ds-11759.

> The bubble allows visitors to have a clear view of the equipment without
> allowing them access to the laboratory area.[22]

The public could observe these events, as well as interact with informative
displays and view films projected onto three large screens inside the pavilion. Lundy referred to this as "an interesting experiment in opacity and
transparency."[23]

The AEC Pavilion showcased not only the peacetime possibilities of atomic
energy but also the potential of air-supported structures. In addition to being one
of the earliest architectural demonstrations of pneumatics, it was recognized for
its size, mobility and open plan, as well as its structural and formal complexity.
Enclosing 22,000 square feet, the pavilion's 300-foot-long curvaceous and open-
ended form (with a maximum width of 126 feet and maximum height of 54 feet)
was comprised of two fire-resistant vinyl-coated nylon skins, with a 4-foot thick
air-filled insulating envelope separated into eight compartments.[24] The 28-ton
traveling pavilion, of which the structural fabric was less than 6 tons, could con-
veniently be packed into the size of a standard railroad box car (5,000 cubic feet)

for shipping.[25] It was also economical ($4.50/square foot + $25,000 site work) and could be erected in only three to four days by 12 unskilled laborers.[26]

Completely supported by air pressure and powered by two large blowers, the AEC Pavilion "consisted of low pressure cellular dual-wall end sections with the center section being of a double walled air-supported construction."[27] A 4-foot deep air space between the inner and outer membrane allowed the air pressure of the interior and exterior skins to be individually controlled, while also providing additional thermal and acoustical insulation. Featuring two dome-like forms, connected by a narrow mid-section, the structure included air-locked revolving doors set into arched openings at both ends. While the outer skin was pressurized at 0.054 per square inch above atmospheric pressure, the inner skin was maintained at 0.07 pounds per square inch.[28] Apparently, this pressure difference was not perceptible to occupants.[29]

In addition to an unconventional building form and system, the AEC Pavilion provided the maximum safety for its occupants. Designed to withstand a 70 mph gale wind or gusts up to 90 mph, the pneumatic envelope was anchored to a concrete slab foundation at each site.[30] Not only was it fire-resistant, but the absence of conventional rigid structural members (e.g., beams, columns, etc.) also eliminated hazards and created an unobstructed view for the audience. In the event of a tear in one or more locations in the pneumatic membrane, its compartmental structure was designed such that the pavilion would not collapse. In the unlikely, but possible, event that the building envelope sustained extreme damage or pressure loss, it would take at least 30 minutes for complete deflation.[31]

The snowy white exterior of the AEC Pavilion was intended to reflect solar radiation, whereas its matte black interior was intended for complete light control in the exhibit areas and theatre.[32] Its interior lighting is described by an AEC brochure as "reminiscent of the eerie Cerenkov glow which is seen in pool-type nuclear reactors and gamma irradiation facilities."[33] Illuminated by blue-violet lights positioned behind a serpentine system of portable wall partitions, an intentional distance was maintained between occupants and the pavilion's inner-skin. Although an unconventional design, the pneumatic enclosure's dramatic shape was intentionally visually suppressed, so as to not compete with the *Atoms for Peace* exhibition, which was also designed by Lundy. Acoustical performance—typically a challenge in air structures—was enhanced by the variable geometry of the pneumatic enclosure. That is, rather than creating two perfect domes within its interior, Lundy and Bird intentionally warped the shapes to avoid acoustical reverberation.

Although the AEC had previously utilized small portable trailers in the United States as a means to communicate their objectives with the general public, the traveling AEC Pavilion was a considerably larger structure with a more ambitious agenda.[34] As Woods explains:

> From the project's very beginning, the enclosure had been a problem. We intended to move the presentation from city to city in Latin America, but

trailers and rented spaces (the two solutions that seemed the most possible within our limited budget), placed too many limitations on design. I began, therefore, to look into the available structural systems and portable buildings. These early investigations turned up, among other things, a portable air-supported missile maintenance enclosure, known as the Pentadome, which had been developed by the Army for the Redstone Base in Alabama.[35]

Coincidentally, the designer of said Pentadome was Walter Bird.[36] In addition to conducting research on portable buildings, Woods obtained preliminary proposals from a variety of air structure suppliers to evaluate their potential.[37] Given the AEC's needs and budget, the consensus was that a pneumatic enclosure could be an advisable option.

The next step was to hire an architect, who in addition to assisting with the design of the pavilion and selection of a building system would be tasked with designing both the enclosure and its exhibits.[38] The selection of Lundy, a New York-based architect, was based on "his ability to incorporate diverse requirements into a simple solution with the immediate visual impact so important to an exhibition."[39] Lundy, with whom Woods shared his preliminary portable structures research, advised that, given the program requirements and schedule, they move forward with pursuing an air-supported structure. Soon thereafter, they selected Birdair Structures of Buffalo, New York, as the fabricator.

When Lundy approached Bird in 1960 to inquire about the possibilities of a collaboration for the AEC Pavilion, the architect had a strict set of criteria informing his selection of a structural system: he wanted it to be malleable, lightweight, mobile, safe, durable, economical, and able to be fabricated quickly. As Lundy explains: "My responsibility was to satisfy the ACE's [sic] requirements for portability, safety, and low cost. But beyond this, I wanted to do something really significant in exhibit structure design."[40] Lundy initially considered a number of different structural systems—wood structures, steel lamella roofs, aluminum (geodesic-inspired) domes and aluminum tents—ultimately deciding on a dual pneumatic envelope to improve thermal, acoustic and lighting properties. The lightness (minimum weight and bulk) of the pneumatic structure meant that it would be not only more portable but also less labor-intensive to erect, dismantle and move from site to site. After the AEC Pavilion was created, Lundy stated that he "wanted it to be a sassy, unafraid example of US ingenuity."[41]

The AEC Pavilion was very economical in terms of its material usage and labor (Figure 4.4). The original cost of the structure was $99,870, and its extensive exhibits, which Lundy also designed, ran another $70,000.[42] Site work for the earth anchors or concrete slab varied at each location, but typically cost $25,000, depending on the existing conditions and local labor costs.[43] It typically took three to four days to erect, which according to Lundy was "1/5 to 1/20th the time required to assemble the other building types considered."[44] Because most of the structure was assembled on the ground, it required minimal labor and

FIGURE 4.4   Victor A. Lundy (Victor A. Lundy and others outside the inflatable pavilion he designed for the US Atomic Energy Commission traveling exhibit
"Atoms for Peace," Buenos Aires, Argentina), c. 1960.

Source: Courtesy of Library of Congress Prints & Photographs Division, LC-DIG-ds-13558.

expertise (it could be assembled by 12 untrained laborers). Once the fabric was
attached to the two rigid end frames with built-in revolving doors (the only task
not carried out at ground level), the structure only took 30 minutes to inflate. By
creating a prefabricated, kit-of-parts air structure, Lundy and Bird demonstrated
not only the technical and sculptural potential of pneumatics but likewise their
value as an affordable and expedient construction type.

   Although a general contractor was brought on board at each location to coor-
dinate assembly and installation, due to its uncommon building construction,
the AEC Pavilion required "constant personal supervision in all phases of the
work."[45] Despite its relatively quick assembly time and comparative ease of mobility, the prefabricated structure and its exhibition elements presented what Woods
describes as "a sizable problem in logistics."[46] He adds:

> But the success of the exhibition so far more than makes up for the dif-
> ficulties. The first three weeks of the exhibit proved the tremendous initial
> impact created by the sensational nature of the exhibit coupled with the
> architect's bold and sensitive solution, which is unquestionably the most
> successful exhibition structure I have seen. [Figure 4.5][47]

FIGURE 4.5 Victor A. Lundy, air-supported exhibition building for the US Atomic Energy Commission, Buenos Aires, 1960.

Source: Courtesy of Victor A. Lundy Archive, Library of Congress, Prints & Photographs Division, LC-DIG-ds-11012.

Prior to the AEC Pavilion, Lundy had never worked with pneumatics. In addition to meeting the AEC's requirements that the structure be mobile, safe and affordable, the architect was intrigued by the formal challenges and potentials of this novel construction type. According to Lundy,

> As an architect used to the sense of control over his medium, I was aware in my first direct experience with air structures that the pneumatic structure was trying to control me. But I chose it to solve the AEC Exhibit Building because of the challenge and opportunity it offered as a malleable sculptural device to enclose the spatial requirements.[48]

Adding that the pavilion was "a first application of air-support principles to a building of true architectural pretension,"[49] Lundy attributes the form to an expression of programmatic requirements—that is, the physical form communicated the logic of its interior plan.[50]

Soon after the AEC Pavilion premiered in South America in 1960, *Architectural Forum* acknowledged it as "a great balloon for peaceful atoms."[51] That same year, Lundy received a Gold Medal for the project at the Buenos Aires Sesquicentennial International Exhibition, as well as a Silver Medal for Engineering from the

Architectural League of New York in 1965. Lundy and Bird's AEC Pavilion was published in a variety of magazines and journals, including *Americas, Architectural Forum, BRAB Building Research Institute Journal, Nuestra arquitectura, Metals & Controls, Architectural Forum* and *Time Magazine.* Although it is unclear as to how many people entered the AEC Pavilion during its lifespan, it is known that while in Rio de Janeiro, it welcomed over 200,000 visitors.[52]

In 1968, the British architectural critic and historian Reyner Banham— an early advocate for inflatables who was originally trained as an aeronautical engineer—proclaimed the project to be "the first great monument of environmental wind-baggery."[53] According to Banham, what set the AEC Pavilion apart from other inflatables was both its size and formal complexity. He writes, "the result remains virtually the only air structure to date with any pretensions to architectural sophistication."[54]

Although not an entirely new building type at the time, the AEC Pavilion adopted the technological prowess of the military-industrial complex and applied this knowledge towards the advancement of pneumatic architecture. In addition to satisfying a range of specifications put forth by the AEC, including a solution that was portable, safe and low cost, Lundy's objective was to "make a major break-through in exhibit structure design."[55] This first pneumatic collaboration with Bird demonstrated on a global scale not only the flexibility and economic efficiency of air structures, but likewise celebrated their promise as a new outlet for architectural expression.[56]

Brass Rail Pavilions

In 1964, Lundy and Bird designed and fabricated ten pneumatic structures for the New York World's Fair at Flushing Meadows Park in Queens, New York (Figure 4.6). Operating as snack and refreshment stands for two consecutive seasons (April–October), the award-winning Brass Rail Pavilions further demonstrated both the technical and sculptural possibilities of this novel construction type. Distributed throughout the fairgrounds, the ten "Air Flowers" were comprised of white fiberglass fabric, inflated and maintained at a low internal constant air pressure.[57] Each of the ten pavilions was covered by a fiberglass canopy, which acted as a podium for the inflated forms above. A large cutout in the center of the canopy, through which a large central mast rose to support the pneumatic roof, also allowed a view of the air structures from below.[58] Underneath their signature roofs, which also protected customers from the rain, fair goers would find a central refreshment center and two sales counters, as well as a terrace, seating area and restrooms.

Rising some 75 feet, and 60 feet in diameter, the "Air Flowers" could easily be spotted from a distance. At night, strong lights mounted on their masts illuminated the pneumatic pavilions from within. Resulting in a network of glowing forms, the refreshment stands also served as beacons and way-finding devices. Describing them as "architectural foliage," Lundy was intent on ensuring that

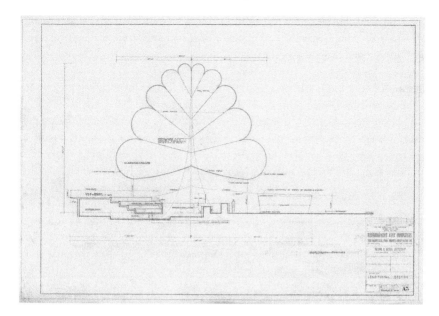

FIGURE 4.6   Victor A. Lundy, refreshment stand, 1964–65 New York World's Fair, for the Brass Rail Food Service Organization, Inc., longitudinal section.

Source: Courtesy of Victor A. Lundy Archive, Library of Congress, Prints & Photographs Division, LC-DIG-ds-11757.

their visual identity be easily understood in contrast to the fair's other pavilions.[59] Although Lundy began his design for the pneumatic structure by working with a clay model, it was the properties of inflating fiberglass fabric that excited him most.[60] Working closely with Birdair, Inc., to maximize efficiency by reducing the number of curved sections and fabric size, Lundy was also adamant about accentuating the vertical striping of the pneumatic "petals." The architect writes, "I remember feeling AEC was too bland, too plain. The vertical lines make it look more like a growing thing."[61]

Coincidentally, the 1964 World's Fair was the first time that New York City waived the 20-year longevity-building requirement for its fair structures.[62] This provided a unique opportunity for architects to design pavilions employing new and innovative materials—such as plastics and fiberglass—that might otherwise have been deemed "too temporary." In response to the possibilities afforded by this longevity waiver, Lundy once again turned to pneumatics as the ideal structural solution to efficiently, economically and experimentally execute his sculptural visions. As Lundy explains,

> Surrounded by purposeful and expensive structure executed in steel and concrete which can stand forever and which had larger budgets, I think

this demonstrates how it is possible through twentieth century technology to create huge volumes of architectural form and space through relatively simple and inexpensive means.[63]

Although rather modest in budget, the impact of Lundy and Bird's contributions to the New York World's Fair did not go unnoticed.[64]

In her review of the fair in April 1964, *New York Times* architectural critic Ada Louise Huxtable identified their ten pneumatic refreshment pavilions as one of two exceptions to what she otherwise refers to as "fascinating mediocrity" and "trick-or-treat architecture—a full range of the current jazzier clichés."[65] She writes:

> The Brass Rail's inflated white balloon-flower canopies by Victor Lundy, based on an experimental "aero-structure" design, do what was not done officially: Spotted about the fair they unify the scene by their repeated grace notes, cloudlike in daylight, glowing at night.[66]

The ten restroom and refreshment stands not only functioned as programmatic infrastructure to support the needs of fair goers but also brought visual delight.

Supported by a steel mast designed to withstand a calculate drag load of 13,500 pounds at a point 30 feet above the ground, each of the ten Brass Rail Pavilions took over a 60 × 100-foot site (Figure 4.7). Celebrating their literal and conceptual lightness, Lundy's intent was to produce the effect of floating.[67] Below each suspended "Air Flower," its various elements, curved to reflect the forms above, were painted in five different shades of bright blue. The cables and steel columns were painted gold, highlighting their contrast against that of the white fabric. In addition to their practical and efficient design, Lundy was determined to maximize their visual impact and capacity for architectural expression.[68] The architect explains, "whatever it may suggest to different people, there is no denying it is a brave roof over a rather simple function."[69]

Although Lundy experimented with lighting in the AEC Pavilion, the Brass Rail Pavilions provided an opportunity to further consider the possibilities of both natural and artificial lighting effects. According to Lundy, "the pneumatic structure is a liberated, free vehicle for creating space and form, for controlling light."[70] The translucent and transparent qualities of fabric allowed the architect to explore the patterning and stitching of the fiberglass forms during the day, as well as at night. Rather than simply creating a bubble or dome-like form, Lundy asserted artistic control over the pneumatic envelopes. By day, the network of ten pavilions was easily identified by fairgoers as a cluster of opaque white balloons.[71] At night, their translucent skin was illuminated, casting a reddish glow that rendered them akin to that of a giant raspberry (Figure 4.8). Building off of Bird's expertise with air structures, Lundy asserted his personal and professional stance that the duty of the architect is not only to enclose space, but also to sculpt it.[72]

FIGURE 4.7   Victor A. Lundy, refreshment stand, 1964–65 New York World's Fair, for the Brass Rail Food Service Organization, Inc.

Source: Courtesy of Victor A. Lundy Archive, Library of Congress, Prints & Photographs Division, LC-DIG-ds-11761.

Not unlike the AEC Pavilion, at the World's Fair Lundy and Bird delivered a design that successfully fulfilled a variety of specific programmatic needs. Lundy describes how the Brass Rail Pavilions were designed to maximize both their functional and "fun" potential:

The problem assigned to the Architect was to do something interesting with the basic refreshment services for the great mass of people at the Fair. These are stands selling food items, housing rest room facilities, etc. My solution to this rather pedestrian problem was to take a simple thing and do something sassy and spectacular with it in a dimension large enough to be really seen and appreciated at the Fair. What I have come up with is

FIGURE 4.8 Victor A. Lundy, refreshment stand, 1964–65 New York World's Fair, for the Brass Rail Food Service Organization, Inc.

Source: Courtesy of Victor A. Lundy Archive, Library of Congress, Prints & Photographs Division, LC-DIG-ds-11083.

something that has been done in a real spirit of fun—it has to do with a fair, carnival atmosphere.[73]

Despite working within the constraints of a limited budget and timeframe, they not only were able to solve a complex problem, but also offered a solution that demonstrated the previously untapped architectural potential of pneumatic enclosures. Lundy adds, "I think if this project has significance it is simply that with very modest means of cost and budget and problem, something rather grand was created."[74] In 1964, only days after the fair opened, architect Philip Johnson called Lundy to ask: "What have you done? We are all wild about them!"[75]

The Promise of Pneumatic Architecture

Lundy and Bird's two pneumatic projects index a temporal juncture in the transformation of architectural modernism—one that privileged not only the flexible, mobile and near-instantaneous qualities of inflatables, but likewise the appropriation of aeronautical engineering technologies to advance the design and construction of demountable structures in the mid-twentieth century. In lieu of prevailing scholarship on the history of pneumatics, which typically privileges the "trippy, cheap, light" ethos[76]—or structural instability of air structures, propped up by the artistic and architectural counterculture of 1960s and 1970s—these collaborations between Lundy and Bird evidence a key moment in the history of building construction. As impermanent yet durable constructions, both the New York World's Fair and AEC pavilions demonstrated how air structures could offer both an economical and formally novel alternative to conventional construction materials and methods. Yet, despite the architect's claim at the 1st International Colloquium on Pneumatic Structures in Stuttgart, Germany, in 1967 that pneumatics "will revolutionize the construction of the future"—a statement reaffirmed by many of his peers at the time—the AEC Pavilion and Brass Rail Refreshment Stands would be the only two pneumatics projects Lundy and Bird would complete together.[77]

Importantly, the technical and formal innovation indexed by Lundy and Bird's inflatable pavilions set the stage for the rise in popularity and acceptance of air structures. For example, almost a decade after the completion of the AEC Pavilion, the 1970 World Expo in Osaka, Japan, featured nearly a dozen different pneumatic structures, including the long-span cable-stiffened dome of the US Pavilion and the air-filled arches of the Fuji Group Pavilion.[78] The following year, the countercultural architectural collective Ant Farm published *Inflatocookbook* (1971), which became (and has remained) the go-to do-it-yourself (DIY) manual for pneumatic experimentation.[79] Likewise, architect Cedric Price and engineer Frank Newby published *Air Structures: A Survey* (1971), an extensive research report covering the history, principles, applications and technical specifications for air-filled enclosures.[80] Despite being a rather short-lived architectural phenomenon, which appeared to fizzle out by the mid-1970s, the contributions of Lundy and Bird shed light on the promise of pneumatic architecture.

Notes

1. V. A. Lundy, "Architectural and Sculptural Aspects of Pneumatics," in *Proceedings of the First International Colloquium on Pneumatic Structures*, May 11–12, 1967, University of Stuttgart, Germany, ed. Frei Otto (Stuttgart: International Association of Shell Structures, 1967), 11. For Walter Bird's paper, see W. W. Bird, "The Development of Pneumatic Structures, Past, Present and Future," in *Proceedings of the First International Colloquium on Pneumatic Structures*, May 11–12, 1967, University of Stuttgart, Germany, ed. Frei Otto (Stuttgart: International Association of Shell Structures, 1967), 1–9.

2. Lundy, "Architectural and Sculptural Aspects of Pneumatics," 11.
3. "Man must control and guide his opportunities for shelter and structure, purposefully, with resolution and with his art, into poetic, brilliant forms, making spaces of appropriateness and high aspiration; so that finally air supported structures become true architecture, become sculptural containers, beautifully and purposefully proportioned over man's spaces—directed by the artist-architect—not just great bubbles or Quonset huts over swimming pools." Lundy, "Architectural and Sculptural Aspects of Pneumatics," 16.
4. Lundy, "Architectural and Sculptural Aspects of Pneumatics," 16.
5. Bird, "The Development of Pneumatic Structures, Past, Present and Future," 4. See also Walter Bird, "Air Structures," *Building Research* 9, no. 1 (January/March 1972) (Building Research Institute): 6–7.
6. Roger Dent, *Principles of Pneumatic Architecture* (London: Architectural Press, 1971), 34–35.
7. According to Bird, "In 1956, approximately ten years after I developed the first air-supported radome, we organized Birdair Structures, Inc. to carry out research, development and fabrication of lightweight portable structures for commercial as well as military applications." Bird, "The Development of Pneumatic Structures, Past, Present and Future," 2. For further information on Walter Bird's early pneumatic works, see Annette LeCuyer, *ETFE: Technology and Design* (Basel: Birkhäuser, 2008), 21–22.
8. "From humble beginnings of developing early radomes and rapid deployment command shelters, Bird and his team went on to develop commercial applications in bulk storage and removable sports facility covers. As a result, in 1957, the Buffalo, N.Y., home of company founder Walter Bird was pictured on the front cover of LIFE Magazine. It featured an air-supported pool enclosure in winter, as a glimpse toward life in the future!" Accessed February 13, 2018, www.birdair.com/about/company-history.
9. Walter Bird: "The first commercial applications of air structures came about 1956 when warehouses and pool enclosures were designed and constructed. But the air structure remained a novelty to many. Building codes, which were established around conventional building, did not cover air structures and special arrangements had to be made for their construction. As the air structure industry was new, there were no recognized design standards or regulations. [. . .] Most problems resulted not from inherent weaknesses in the air structures, but from poor design." Bird, "Air Structures," 6.
10. Bird, "The Development of Pneumatic Structures, Past, Present and Future," 1.
11. Ibid., 2–3.
12. Bird, "Air Structures," 7.
13. Lundy, "Architectural and Sculptural Aspects of Pneumatics," 16.
14. "The truth of the pneumatic structure Lundy is its alive straining to fill out, to escape, to burst through the thin skin containing it, restraining it. It wants to become big round, bubbular shapes." Lundy, "Architectural and Sculptural Aspects of Pneumatics," 15.
15. "Architecture is sculpture as well as enclosure of spaces." Lundy, "Architectural and Sculptural Aspects of Pneumatics," 16.
16. The Atomic Energy Commission (AEC) was formed as a result of the Atomic Energy Act of 1946 during the Harry S. Truman administration (1945–53), and continued to be developed under the presidency of Dwight D. Eisenhower from 1953 to 1961, before it was ultimately dismantled in 1975.
17. Lundy, "Architectural and Sculptural Aspects of Pneumatics," 11.
18. United States Atomic Energy Commission, *Atoms at Work/Atomos en Accion* (Washington, DC: U.S. Government Printing Office, 1963), 11.
19. "In the center of the floor, a railing separates the general audience from the technical facilities, which consist of a gamma irradiation pool built and operated by Brookhaven

Laboratory, a portable research and training reactor, designed and operated by Lockheed, and a reactor simulator furnished by Leeds and Northrup. The reactor and reactor-simulator are enclosed by an air-supported clear plastic hemisphere with insulation material adhered to its upper side to provide a quiet working area for the scientists." Albert H. Woods, "Atoms for South America," *Industrial Design* 8, no. 1 (January 1961): 64.

20. United States Atomic Energy Commission, *Atoms at Work/Atomos en Accion*, 11.
21. Woods, "Atoms for South America," 61.
22. Lundy, "Architectural and Sculptural Aspects of Pneumatics," 14.
23. Ibid., 14.
24. Ibid., 13.
25. The AEC Pavilion weighs only 28 tons, which includes all hardware and fittings.
26. David Allison, "A Great Balloon for Peaceful Atoms: Architect Victor Lundy Shapes an Air-Supported Sculpture for the A.E.C.'s South American Exhibit," *Architectural Forum* 113, no. 5 (November 1960): 142–45, 204.
27. Bird, "The Development of Pneumatic Structures, Past, Present and Future," 4.
28. For a closer reading of the AEC Pavilion with respect to "atmosphere," see Susanneh Bieber, "Atmospheric Pressures: Victor Lundy's AEC Pavilion and the Sociopolitical Climates of Inflatable Architecture," *Journal of Architectural Education* 73, no. 1 (2019): 32–45.
29. Woods, "Atoms for South America," 64.
30. *Atoms at Work/Atomos en Accion*, 13–14.
31. Ibid., 14.
32. Woods, "Atoms for South America," 64.
33. *Atoms at Work/Atomos en Accion*, 14.
34. It should also be noted that the AEC overseas exhibit program hosted a number of presentations and smaller scale exhibits on the peacetime uses of atomic energy from 1955 to 1962. See *Atoms at Work*, 10.
35. Woods, "Atoms for South America," 61.
36. Designed and manufactured for the US Army by Birdair Structures, Inc., the "Pentadome" was a complex of five pneumatic domes, whose floor area was in excess of 50,000 square feet. At 85 feet tall, the center dome featured a 150-foot clear span, while the adjacent four domes were comprised of 100-foot spans. The pneumatic complex was used for the assembly of missiles, antenna construction, and military exhibits. See Benjamin H. Evans, "Air Structures Forum," *Building Research* (April/June 1971): 8.
37. Woods, "Atoms for South America," 61–62.
38. Ibid., 62.
39. Ibid.
40. Victor Lundy, "AEC Exhibit," *Building Research* 9, no. 1 (January/March 1972) (Building Research Institute): 47.
41. According to architect and Lundy scholar Donna Kacmar, Lundy stated this in an "AIA award entry text." See Donna Kacmar, ed., *Victor Lundy: Artist Architect* (New York: Princeton Architectural Press, 2019), 199, 225, n. 15.
42. It should be noted that in his CV, Lundy lists the project as costing $1,000,000. Victor Lundy, CV, eight pages typed, undated. Victor A. Lundy Archive, Library of Congress, Washington, DC.
43. Allison, "A Great Balloon for Peaceful Atoms," 142–45.
44. Lundy, "Architectural and Sculptural Aspects of Pneumatics," 13.
45. "Even here, because of the prefabricated nature of the exhibition and the fact that it differed in many respects from normal building construction, we had to provide constant personal supervision in all phases of the work." Woods, "Atoms for South America," 64–65.

46. "The prefabrication of a 22,000-square-foot exhibition including its enclosure and its movement to separate locations in foreign countries presents a sizable problem in logistics." Woods, "Atoms for South America," 65.
47. Woods, "Atoms for South America," 65.
48. Lundy, "Architectural and Sculptural Aspects of Pneumatics," 12.
49. As Lundy explains "the air structure became a total, integrated, malleable shape molded to house a theatre and a technical center." Lundy, "Architectural and Sculptural Aspects of Pneumatics," 11.
50. Lundy, "AEC Exhibit," 48.
51. Allison, "A Great Balloon for Peaceful Atoms," 142.
52. "Atoms at Work: M&C Participates in the United States Atomic Energy's Atoms for Peace Program," *Metals & Controls*, 5.
53. Reyner Banham, "Monumental Windbags," April 18, 1968. Reprinted in Marc Dessauce, ed., *The Inflatable Moment: Pneumatics and Protest in '68,* 31. Requoted in LeCuyer, *ETFE: Technology and Design*, 22.
54. A year later, Banham highlights the project in his 1969 book *The Architecture of the Well-tempered Environment*. He writes: "Besides its durability, it is notable among inflatable structures for its size, complexity and open form on plan. Whereas most air-supported structures tend to be simple domes, or elongations of domical forms that still retain a closed figure in plan, the AEC pavilion is better described as an open-ended vault, or half-tube, deformed to produce two approximately hemispherical spaces joined by a central neck, and entered by means of arched porches, about the same diameter as the neck, at either end. Internally, there is a small inflatable dome to house a model atomic reactor, and sundry rigid, non-inflatable partitions, projection screens, and so forth. The precise distribution of credit for the design between Lundy, Bird and the consulting engineers, is not easy to fix, but the result remains virtually the only air structure to date with any pretensions to architectural sophistication." Reyner Banham, *The Architecture of the Well-Tempered Environment* (Chicago, IL: University of Chicago Press; London: The Architectural Press, 1969), 270–72.
55. Lundy, "Architectural and Sculptural Aspects of Pneumatics," 11.
56. According to Lundy in 1972, "It wears well and is as new today as it was when it was built; it can be called a purposeful work of architecture, with spaces proportioned, created, and lighted for special purposes and effects. It is when one understands the essence of this concept and the fact that this medium can be manipulated into mighty feats of architectural enclosure, that it gets exciting." Lundy, "AEC Exhibit," 47.
57. Lundy repeatedly refers to the World's Fair pavilions as "Air Flowers," whereas in some instances they are called "Space Flowers." See Victor A. Lundy, "Refreshment Complexes for the Brass Rail New York World's Fair, 1964–65," undated, 2.
58. According to Lundy, "The entire site of 60 x 100 feet is roofed over with an interesting canopy which acts as a white podium for the white air sculpture, which is also of white fiberglass. It rises and is attached to the central mast which supports the air structure and has a large cutout in the center allowing a view of the air structure from below." Lundy, "Refreshments Complexes for the Brass Rail New York World's Fair, 1964–65," undated, n.p. A three-paged typed statement by the architect describing the Brass Rail project. Victor A. Lundy Archive, Library of Congress, Washington, DC.
59. According to Lundy, "One of the realities of this problem was that these Refreshment Centers are scattered at random throughout the Fairgrounds. It seemed like an excellent opportunity, with the great variety of forms and shapes and solutions at the New York World's Fair, to use the fact that there are many of these units, to full advantage as a unifying recurring visual image of the Fair as a whole. The brave, sassy shape of the air structure is intended simply as an abstract, visual image that people will identify readily through repetition throughout the Fair. It was intended in a sense

as 'architectural foliage.' It was purposely made to a dimension big enough to be seen." Lundy, "Refreshments Complexes for the Brass Rail, New York World's Fair, 1964–65," n.p.

60. As Lundy explains, "there seem to be endless possibilities for taking advantage of combinations of materials, opacity, translucency, transparency of fabrics, patterns of stitching and joints; effects through these combinations, of natural light coming through in controlled ways by day; and the reverse effect by night." Lundy, "Architectural and Sculptural Aspects of Pneumatics," 16.

61. Victor Lundy, sketch for the Brass Rail Pavilion, dated March 9, 1963. Victor A. Lundy Archive, Library of Congress, Washington, DC.

62. As *New York Times* reporter John M. Lee explained in 1964, "Owens-Corning Fiberglass estimates that about a million pounds of glass fiber are employed in about two dozen fair pavilions. The material is used in everything from telephone booths to airborne roofs at the Brass Rail refreshment centers." Importantly, Lee adds, "the opportunity for the use of plastics and synthetic fibers at the fair came when New York City waived the 20-year longevity requirement of the building code for fair structures." John M. Lee, "Plastics Abound in Fair Buildings: Industry Seeks to Capitalize on Construction Showcase," *NYT*, April 26, 1964; ProQuest Historical Newspapers: *NYT*, 85.

63. Lundy, "Refreshments Complexes for the Brass Rail, New York World's Fair, 1964–65," n.p.

64. Lundy indicates in his CV that the ten pavilions cost a total of $1 million. Victor Lundy, CV, eight pages typed, undated. Victor A. Lundy Archive, Library of Congress, Washington, DC.

65. Ada Louise Huxtable, "Architecture: Chaos of Good, Bad and Joyful," *NYT*, April 22, 1964; ProQuest Historical Newspapers: *NYT*, 25.

66. Huxtable, "Architecture: Chaos of Good, Bad and Joyful," 25.

67. "All of the elements below are painted in gay curved lines in five shades of bright blue which echo the planned cusped curves of the air structure above. From a distance this will look like a piece of white sculpture sitting on a white podium that floats over the ground. The supporting steel columns and cables are painted a rich gold against the white of the fabric." Lundy, "Refreshment Complexes for the Brass Rail New York World's Fair, 1964–65," n.p.

68. According to Kacmar, "Lundy Designed the Restrooms and Public Counters for Ordering, and the Tables, Chairs, and Signage," in Kacmar, *Victor Lundy*, 209.

69. Lundy, "Refreshment Complexes for the Brass Rail New York World's Fair, 1964–65," undated, 2.

70. Lundy, "Architectural and Sculptural Aspects of Pneumatics," 17.

71. "Well, thank God for The Brass Rails. If you have a child in tow, these innumerable havens under their clusters of white balloons are the real answer. By now you know you can't tempt a child with turbot blazed in Pernod or salmon flown in especially—no, his closed, adamant mind wants a hot dog or a hamburger, and now. The Brass Rails have them, and tiny pizzas too, along with those sweet sweet drinks of suspicious colours which are the equivalent of blood plasma to the young. (And if your little boy asks you about a little boys' room, The Brass Rail has them too.)" "Eating at the Fair, N.Y.," *Vogue* (July 1964).

72. "Architecture is sculpture as well as enclosure of spaces. The great works of architecture work both ways, from the outside and from the inside. I see air supported structures as a great new vehicle of making giant architectural sculpture and space for man—with big, brilliant strokes." Lundy, "Architectural and Sculptural Aspects of Pneumatics," 16.

73. Lundy, "Refreshments Complexes for the Brass Rail, New York World's Fair, 1964–65," n.p.

74. Ibid.
75. Kacmar, *Victor Lundy*, 209, 225, n. 23.
76. In a 1971 edition of Stewart Brand's infamous *Whole Earth Catalog*—a grassroots, DIY, countercultural magazine—Brand questioned the viability of pneumatics, stating: "Inflatables are trippy, cheap, light, imaginative space, not architecture at all. They're terrible to work in. The blazing redundant surfaces disorient; one wallows in space. When the sun goes behind a cloud you cease cooking and immediately start freezing. Environmentally, what an inflatable is best at is protecting you from a gentle rain. Wind wants to take the structure with it across the country, so you get into heavy anchoring operations." *The Last Whole Earth Catalog* (Portola Institute, Inc., 1971), 107.
77. "In this sense, pneumatic structures offer great possibilities in the hands of creative architects and engineers to make new forms or wonder and possibility that are truly of our time. Pneumatic structures are of this age and into the future." Lundy, "Architectural and Sculptural Aspects of Pneumatics," 11.
78. According to the *Expo '70 Official Guide,* the United States Pavilion featured an "elliptical translucent domed roof" and was, at the time of its completion, "the largest and lightest clear span, air supported roof ever built." Similarly, the Fuji Group Pavilion was "the world's largest pneumatic structure" to date. For a comprehensive overview of the 1970 Osaka Expo, see *Expo '70 Official Guide* (Osaka: Japan Association, 1970), 57 and 191.
79. Ant Farm, *Inflatocookbook*, 1971/73.
80. Cedric Price, Frank Newby and Robert H. Suan, *Air Structures: A Survey* (London: Her Majesty's Stationery Office, 1971).

5

SAARINEN'S SHELLS

The Evolving Influence of Engineering and Construction

Rob Whitehead

Collaborations and Compromises

Over the span of a decade, the architects of Eero Saarinen & Associates collaborated with the structural engineers of Ammann & Whitney in the design of three very different, unprecedented, long-span concrete shell-like projects with high levels of technical complexity: Kresge Auditorium (1951–55), TWA Terminal (1956–62) and Dulles Airport Terminal (1958–63). Saarinen initially challenged the traditional notion that structural shell performance and building expression were indelibly linked. As a result, they endured burdensome challenges of analyzing, documenting and constructing these anomalous projects.

The first stages of this process resulted in more compromise than collaboration between Saarinen and Ammann & Whitney, in part because the building forms weren't intended to reflect optimized engineering-based solutions. However, for each successive project, the priorities evolved until a shared set of goals developed between the two firms, which eventually resulted in a thoroughly documented, technically innovative and formally expressive solution for Dulles Terminal (Figure 5.1).

The differences in the projects, in both building form and performance, can be explained in part by examining the evolving nature of the collaborative design relationship between the participating firms. As N. Keith Scott observed at the conclusion of Kresge, "in many cases (Saarinen) has relied upon the sheer ingenuity of modern technology to get him out of difficulties that would have presented insurmountable obstacles a quarter of a century ago."[1] Some improvements in the ensuing thin-shell structures were necessitated by the significant problems that first incurred at Kresge Auditorium.

FIGURE 5.1 The three shell projects—(*top* to *bottom*) Kresge Auditorium, TWA Terminal and Dulles Terminal—show an evolving shift towards structural and construction considerations across a decade of collaboration.

Source: Photos by Balthazar Korab. US Library of Congress, Balthazar Korab Archive.

Kresge Auditorium (1951–55)

The Kresge Auditorium on MIT's campus was designed to be the nation's largest free-standing concrete shell upon completion (160 ft. span (48.8 m)) with a thickness that was proportionately thinner than an egg. The shell's unique geometry—one-eighth of a sphere which rested on three points—was visually striking. According to Saarinen, it was intended to be both progressive and contextual: "The strongest, most economical way of covering an area with concrete is with a dome and a dome of thin-shell concrete seemed right for a university interested in progressive technology."[2] Yet, unbeknownst to many, the form was structurally speculative and poorly resolved. In 1954, shortly after a portion of the shoring was removed to reveal the shell, *Harper's Bazaar* photographed Saarinen standing in front of the shell;[3] although Saarinen didn't mention it at the time, the building had started deforming past the allowable limits. It was failing.

Once the shoring was removed to allow the shell to stand on its own, the forces moved in ways the designers hadn't anticipated. Interruptions in the shell's geometry, dictated by the triangularly shaped segmented spherical form, transferred stresses from the membrane into the structure's stiffest elements—the three perimeter arches. This produced a large amount of unanticipated bending stress in the arches, which caused them to undergo six weeks of unhindered creep bending and cracking the shell. The sagging was up to 5″ (12.7 cm) in the middle of the arches, which was more than three times the anticipated amount. Scaffolding was placed back under the structure and a plan to permanently add steel support columns underneath the arches, concealing these columns as vertical window mullions, was quietly put in place.[4]

These structural failures, in addition to the difficult and costly construction conditions that proceeded them, bolstered the critical view that the building form was derived from neither functional nor engineering-based logic—both points eventually conceded by Saarinen years later. The problem with the building wasn't *just* the flawed form, it was also the process of design and production that proceeded its construction. The potential for problems began at the onset.

In 1951, before hiring the structural engineers, Saarinen drew sketches of Kresge that proposed various double-curved thin-shell options, including some spherical forms that alluded to the nearby domed roofs on campus.[5] Saarinen's aesthetic aspirations were tempered by three central, often competing challenges related to the project's form: the plan geometry of a functional auditorium, the acoustical consequences of the space's sectional volume and the structural logic that would span long distances and provide enclosure. Saarinen's office built many models, looking for a form that would solve all three conditions. No single form seemed to work.

Saarinen described how he became fond of one model, which was initially rejected by his design team and dubbed the "Vulgar Freak" because of its lack of structural logic. Saarinen disagreed. He thought that the three-pointed dome would match a fan-shaped auditorium plan and would be able to defy the "earth-bound" aesthetics of other domes by allowing large arched windows around the perimeter. He demonstrated the form with a grapefruit peel, renamed the design "The Loved One" and decided that it would be the basis for the final building design.[6]

Although Saarinen stated publicly that the unique form was structurally appropriate, it wasn't. It was unique among other shells because it was based on Euclidian geometry and aesthetics, not structural forces. The top of the shell looked like a sphere, but it didn't have the continuous perimeter support of a typical dome or the continuity of force transfer found in the "arches and hoops" of domes. Unfortunately, the spherical dome curvature wasn't appropriate for shells with concentrated supports, either—other anti-clastic shell forms generated their double-curved geometry from other engineering-based precedents or form-finding techniques.[7] Saarinen hired Ammann & Whitney, the nation's leading expert in the emerging field of concrete shell design to help complete the building, but only *after* he'd established the form (Figure 5.2).

One of the firm's partners assigned to the job, Boyd Anderson was a national award-winning engineer known for his ability to design suspended steel construction and his experience with traditional concrete shells.[8] Anderson was tasked with working with Saarinen—a role he would assume for the next decade as the project engineer for all three projects. According to Saarinen's design partner, Kevin Roche, Anderson developed a reputation with Saarinen's office as an "open-minded collaborator that never dealt in absolutes."[9] Although Anderson would eventually become more involved in preliminary design discussions with Saarinen, the collaboration for Kresge was quite one-sided. Anderson's task was to make the shell form that Saarinen had proposed work structurally.

Roche justified the exclusion of the engineers from the deliberations on the shell's form because "the structural constraints were just one of many issues that needed to be balanced by the project's design."[10] Asking the engineer to "make it work" is understandable in the context of more traditional mid-twentieth-century contractual responsibilities and practices, but concrete shells are different than other projects—the form is *the* primary determinant of behavior. The exclusion of structural consultation in the development of Kresge's form was ill-advised, and it became the project's central liability for all ensuing problems.

The engineering challenge was exceedingly difficult, in part because the overall form didn't comply with any previously tested or constructed shells. Anderson and his team endeavored to make the building work by simply modifying the shell thickness, arch depth and reinforcing schemes of the proposed form. Anderson believed a uniform shell thickness of 3.5 inches (9 cm) would be adequate for the shell surface to resist the loads and force transfer through the surface (i.e., "membrane action"). However, the continuity in shell surface that normally occurs in domes was interrupted in this case by the tripod supports; Anderson suggested

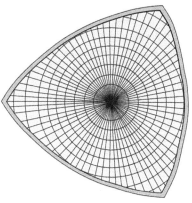

FIGURE 5.2 *Left* to *right*, *top* to *bottom*: Eero Saarinen posing behind a support at Kresge Auditorium for *Harper's Bazaaar (1955)*; a diagram of the formwork used to merge the triangular and spherical geometries; and the view of the auditorium in context of the other domes on MIT's campus.

Source: *Top left*: Image courtesy of Eero Saarinen Collection, Yale University Archives. *Top right*: Illustration by author. *Bottom*: Photo by Balthazar Korab. US Library of Congress, Balthazar Korab Archive.

that the shell be thickened into an edge beam at the perimeter—arching between the three supports—to stiffen the shell surface and to support the inevitable bending stresses that would occur in a non-idealized structural form. Because the engineers were using first principles of design, they decided to treat the three arches as large curved beams, but they didn't have a good way of knowing how much force would be transmitted to each and may have under-sized the reinforcing.

Ultimately, this engineering choice led to the failure of the building to perform structurally as intended.

As the project moved into the construction documentation phase, more challenges arose. Saarinen's team had never documented a shell before. In the drawings and specifications, the design team chose not to prescribe the means of construction, perhaps because it was a "design/bid/build" contract and they wanted to give the bidding contractors options for cost-savings and/or construction operations.[11] This was a significant departure from the established precedent of other shell projects worldwide in which shells were frequently designed in close collaboration with (or designed by) their builders (e.g., Torroja, Nervi, Candela, Dieste, etc.).[12] Instead of documenting assembly methods in the plans or section drawings, or requiring coordinated efforts to do so, the drawings primarily documented the building's geometric form using a system of graphic grids and reference radial points.[13]

The most surprising omissions were found in the project specifications. The "concrete" section of the specifications mostly used standard language about concrete casting, formwork and testing, abdicating most of the responsibility for coordinating this work to the contractor. The builder was granted the ability to determine the "design, fabrication, and erection of the shell falsework and forms, the pouring procedure (one pour or in sections with joints), and the decentering procedure." The specifications also gave the contractor a choice for how to pour the shell—either as a "single continuous operation" or in separate segments with joints.[14] These are incredibly broad expansions of authority to give to a contractor, particularly because these choices can adversely affect structural performance. Granting this much leeway for interpretation to a builder without documenting requisite performance requirements is even riskier when one considers that this was a publicly bid project that was awarded to the lowest bidder—cheaper prices could involve riskier strategies.

When construction began in May 1953, George A. Fuller Company had to sort through the unique complications of construction with only marginal input and instruction from the design team. As a result, they chose to pour the roof in separate segments because a single pour wasn't possible on a project this large. Offering this choice without adding qualifiers was a mistake. From a technical and structural perspective, these options *aren't* equivalent. A shell relies upon the monolithic behavior of the membrane; allowing the shell (and arches) to be poured at separate times with potentially different concrete mixes and adjoined with cold pour "joints" simply invited the types of problems of cracking and massive deflection that eventually occurred.

The builders started separate pours simultaneously at the three base supports and continued moving upwards to the middle of the dome in a series of pours that ultimately took more than *three months* to complete. Having a double-curved surface is always hard to pour (and screed), particularly for a shell with relatively large rebar and minimal thickness (3.5″ uniform thickness), but it was made more difficult by the integrated arch beams—especially at the supports, as evidenced by the construction photos (Figure 5.3).[15] Making matters worse, the roof

FIGURE 5.3   Complications with concrete pouring and the building geometry. *Top*: Three separate pours progressed towards the top—a process that took all winter. *Bottom*: The double-curved surface, integral beam and thin shell made casting particularly challenging.

Source: *Top*: Courtesy of Eero Saarinen Collection, Yale University Archives. *Bottom*: Courtesy of MIT Museum, Kresge Auditorium Photo Archives.

was poured during the coldest months (December 1953–February 1954). Cold weather creates many difficulties for casting concrete since it requires additional attention, and preparation, to how the concrete is mixed, placed and cured to make sure the concrete achieves its target yield strength.

Upon competition, the project had plenty to critique. Kresge was more expensive than anticipated, took longer to build than expected and didn't perform well structurally.[16] Yet, of most of the articles written about it, only a few made passing references to "difficulties," and many excused the cost, and construction issues, as part of the consequences of experimental formal design. The remediation plan, which added columns under the arches, wasn't even mentioned in any major publications.[17] The extent of these problems, and the role that the design and construction team collectively played in creating these issues, didn't fully come to light until 30 years after the building's completion, when engineers from Ammann & Whitney returned to work on a re-roofing and repair project.[18]

It was clear to the project participants that much of the problems on Kresge stemmed from the fact that the building simply wasn't designed with the means of construction and structural performance in mind. One of Ammann & Whitney's founding principals, Charles Whitney, spoke at a Conference on Thin Concrete Shells, hosted by MIT in June of 1954 (at the same time the structural problems were becoming clear); he only mentioned the project once, tersely stating that shells like Kresge have, "no general rules as to their justification."[19] Even Saarinen joined the criticism, writing in 1958 that, "in retrospect, one has to criticize this building . . . we learned that one cannot depend on geometry for the sake of geometry."[20] But it's not clear that Saarinen ever accepted that the problems with Kresge stemmed from not matching a structurally appropriate form. In a speech in 1957, Saarinen admitted "the reason why these (plastic forms) are being built now . . . is really aesthetic and not economic; and we should face that."[21]

Less than a year after Kresge's completion, the two firms would collaborate again on a very different design, this time for a Trans World Airlines (TWA) terminal building in Idlewild, New York. The design process would be somewhat more cohesive, with new challenges that forced a deeper collaboration.

Trans World Airline (TWA) Terminal (1956–62)

The changes in the working relationship with Ammann & Whitney were evident from the start of the TWA design process. Early on, Saarinen invited Anderson to meet with him because Anderson had been "very patient, and gentle with his guidance (on Kresge)."[22] Anderson recalled that "each time (Saarinen) got a new project he would call us specialists to come and sit together with him so that he could probe us for ideas. Sometimes this would go on for days or weeks!"[23] At one critical point in the shell's design development, this openness to design collaboration would pay off for both parties.

Saarinen wanted to "express the drama and wonder of air travel" through a dynamic building form made of concrete. He saw the challenge as a competition between gravity and aesthetics and expressed his desire to "make the vaults, whose compressive forces are always downward, become soaring rather than earthbound."[24] His initial design, sketched on the back of a menu, showed the roof as an undulating elliptical paraboloid shell with two "wings" cantilevering upward and outwards. Although structural logic informed this initial proposal, it wasn't the final design.

Saarinen dictated a memo/manifesto during this process (approximately 1958) in which he challenged the very idea of prescriptive engineering-based forms. He stated, "(the) structural and rational cannot always take precedent when another form proves more beautiful. This is dangerous but I believe true."[25] Unsurprisingly, Saarinen's modified scheme became more sculptural than structural. Roche stated that "(TWA) was more of a structural problem than a structural solution. It was never meant to be a thin shell."[26]

Saarinen presented the Ammann & Whitney team with a model of the roof as a continuous shell surface with four upward bulging quadrants, all bound at the perimeter by a large undulating edge beam. This scheme repeated the same two complications that had plagued Kresge: the continuous roof surface and the stiffened edge. Anderson, and a young assistant engineer named Abba Tor, worked hard to explain the problems to Saarinen and to seek out a mutually beneficial solution. Essentially, it wouldn't act like a shell at all; it would have to be engineered like a long-spanning curved slab, which would make it significantly thicker, thus risking further structural problems, as dead load weight is the controlling design factor for shells. Tor described the difficulty of disagreeing with Saarinen's designs: "You had to kind of argue your way into it. . . (by bringing out) the possibility of creative solutions and compromises."[27]

Anderson and Tor proposed a simple change: if the four bulges could be split apart—separating the continuous slab—they could be cast independently and would be more autonomous. Happily, Saarinen was also contemplating ways to make the shells more gestural and open on the interior with skylights. A common solution was proposed that solved both issues: the roof was changed from one large undulating shell into four separate quadrants, separated by continuous linear skylights. The benefits were immediately obvious (Figure 5.4).

Breaking the roof into four different pieces meant that it no longer behaved like a shell structure—although it hadn't for quite some time—but it also created a clear structural logic with reduced spans. The perimeter was supported at three points: one corner cantilevered up and out, gestural Y-shaped column buttresses were placed at two of the corners and the final corner was conjoined with the other shell quadrants in the middle of the building with a 44-inch (112 cm) thick center-plate slab "keystone." Each portion of the roof shell could now also have a varied thickness, depending on the desired geometry and structural constraints,

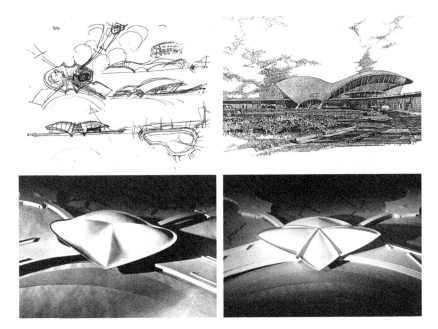

FIGURE 5.4   The evolution of TWA's shell form. *Top left*: Saarinen's initial sketch on a menu. *Top right*: The initial scheme maintained the two "wings" in a relatively viable structural form. *Bottom left*: The "before" model repeated many of the mistakes from Kresge, including a continuous surface and perimeter beam. *Bottom right*: The "after" model integrated skylights along the quadrants and improved the structural and architectural performance.

Source: Courtesy of Eero Saarinen Collection, Yale University Archives. *Bottom*: Courtesy of MIT Museum, Kresge Auditorium Photo Archives.

and each quadrant could ideally be poured in one day, which eliminated concerns for concrete shrinkage and construction joints that occurred at Kresge. Both Tor and Saarinen both take credit for proposing the change.[28]

To develop and document this complicated three-dimensional form into cogent construction documents that improved upon the errors of Kresge required additional collaboration. Saarinen's office turned again to model-making as a way to refine the building's form and to test the building's aesthetics.[29] The only way to accurately build the curvature of these huge models was to generate sectional templates for the different arch profiles of the roof at the edges and the middle of the shells and then to join the pieces together to form a ruled surface with rectangular pieces of paper mimicking the formwork pattern. Roche, who helped build the models, described it as a very methodical process with an underlying logic of 2D to 3D translation that was then reversed to document the form accurately again in 2D for the engineers to complete the construction documents (Figure 5.5).[30]

Ammann & Whitney had no choice but to engineer the building like a piece of large sculpture, relying upon first principles of engineering design. In writing about the TWA engineering process, Anderson and Tor recalled that the geometry of the edge beams was larger than it needed to be and primarily dictated by the architects.[31] Both firms spent a lot of time producing a thoroughly coordinated, and unprecedented set of construction documents, which included construction oversight requirements. They produced more than 130 architectural and structural construction drawings to represent the unique geometric and structural properties of the building's elements.[32] Architectural and structural plans showed contour lines and spot elevations that delineated variations in slab thickness. Drawings also included gridded serial sections with plan and elevational drawings of the columns that described the evolving geometric forms in x, y, and z axes.

Unlike Kresge's specifications, these were more rigorously written to address the unique requirements of the construction operations, including the requirement that the construction process be refined by the design team alongside the contractors, Grove, Shepard, Wilson & Kruge, *before* construction began. One critical measure was related to the pouring process. To ensure better structure behavior, the specifications required that each shell quadrant must be poured continuously in one-day with only minimal allowable dimensional deflection after the scaffolding was removed. Both firms stationed employees on-site permanently during construction. Together, they worked for nine months to prepare a manual

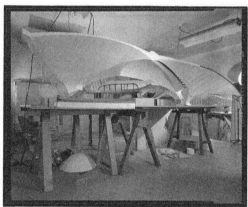 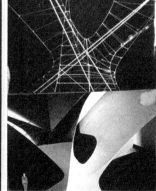

FIGURE 5.5   Large models were used to develop and define the geometry of TWA's form. The ruled surfaces and controlling sectional profiles used to make the models were translated into the plans and sections of the construction documents.

Source: Photos by Balthazar Korab. US Library of Congress, Balthazar Korab Archive.

that outlined the plans for forming, finishing, and testing the tolerances of the concrete.[33]

The support for the roof was formed with a rigorously-controlled grid of scaffolding, upon which the curved wooden formwork was mounted. Computers weren't used to design or engineer the building, but the builder used a computer to help to calculate the *exact* height of each vertical scaffolding post, all 1,800 of them, to make sure the anticipated curvature of the wood setting upon the scaffolding was met within a 1/4" tolerance. After the wooden surface was fully formed and sealed, the contractor spray-painted spot elevations on top side of the formwork (in the same locations as the construction drawings showed) noting the thickness of the concrete at those points as a guide for pouring from above. Three test panels were constructed to simulate the most difficult placing conditions (e.g., angle of incline, crowded reinforcing, etc.) to fully prepare for the pouring process that had been pinpointed with "unforgiving tolerances."

Over 150 workers helped pour each section of the roof; they were given a shirt with a giant number on it that corresponded with an assigned position on the roof. The pouring process was so coordinated that each bucket was color coded with paint to assure that it was being placed in the right location to ensure workability (as mixtures varied). Concrete was placed at a staggering rate of one cubic yard every *two minutes*. Inspection crews of engineers and carpenters, stationed under the formwork and at the ground level below, observed a system of hanging plumbs from the roof. This meant that the next bucket load could be placed at a compensating location for counterbalance if the formwork moved. The largest roof section was 1,000 cubic yards and it took a full 30 hours of continuous labor to pour and finish. To leave the concrete roof visible to planes above, it was coated with 1,500 gallons of silicone waterproofing material that prevented freeze/thaw damage to the concrete facilitating faster runoff of rainwater to keep the roof looking clean (Figure 5.6).

This time, the thoroughness of the design and construction process was rewarded. The finished shell sagged considerably less than anticipated upon removal of the formwork and there were no noticeable shrinkage cracks. The project foreman, Kenneth Morris, gave high praise to the process and participants, calling the effort "the biggest challenge to concrete and concrete men I've seen in my 30 years of construction . . . the teamwork was the finest I've ever seen."[34] Because he died unexpectedly in 1961, Saarinen never saw the building completed. However, he did see the concrete form with the scaffolding removed and boasted in a letter that, although there was a lot of concrete, "it is the least earthbound shell that has ever been built."[35]

Dulles International Airport Terminal (1958–62)

Even though there were meaningful improvements made in the collaborative design efforts for TWA, the team had yet to incorporate the inherent structural

FIGURE 5.6 TWA's construction process was rigorously planned and coordinated. *Left*: Workers wore numbered shirts to correspond with their position on the roof; each quadrant was poured in approximately one day. *Right*: When the scaffolding was removed, the roof sagged less than anticipated and revealed the expressive massing and daylight that defined the aesthetic.

Source: Left: Courtesy of Eero Saarinen Collection, Yale University Archives. *Right*: Photo by Balthazar Korab. From US Library of Congress, Balthazar Korab Archive.

and economic benefits of spatial shell design—topics frequently promoted by the engineering team at Ammann & Whitney.[36] While the firms were preparing construction documents for TWA in 1958, Ammann & Whitney was selected by the US Civil Aeronautics Administration as the prime contractor for the nation's first jet airport terminal in Washington, DC. The engineers, in turn, hired Eero Saarinen & Associates as their architect and the process began again, albeit from a much more experienced perspective on both sides.

Design collaboration on Dulles Terminal was immediate—perhaps as a result of Ammann & Whitney's role as the primary contract holder. Representatives of both firms completed a detailed analysis of existing airports in an effort to understand how the movements of passengers and jets could be optimized to provide a more convenient, flexible and effective set of operations. Ultimately, Saarinen proposed the radical idea to use "mobile lounge" vehicles to transport passengers from the terminal to the jets, allowing the terminal building to be smaller, more efficient and more precisely illustrative of its purpose—the building form was intended to express this compressed movement of passengers through the space from high to low. The design needed to accommodate an eventual expansion, so Saarinen designed the building with a repetitive section that could be easily expanded at each end. This resulted in a large, linear, open room under a sweeping roof with a single-curvature flanked by colonnades on two ends.

Anderson noted that on previous shell projects Saarinen "had absolutely no problem in simply throwing out a design and starting anew," but the range of proposed solutions for Dulles was rather narrow and concentrated.[37] Saarinen developed the building form quickly and assuredly, perhaps as a result of his collaboration with another structural engineer, Fred Severud, on the recently completed Ingalls Hockey Rink at Yale University (1953–58). Saarinen lauded the "structurally effective and beautiful" design of Ingalls and the way that the building expressed the "unique twentieth century technology" of tension.[38] Ultimately for Dulles, Saarinen determined that a similar roof form, one derived from a catenary roof shape, would give a sweeping form with a repeatable section profile (Figure 5.7).

At this point, Saarinen considered Anderson a "trusted collaborator," and Anderson's expertise in long span suspension structures helped to validate the benefits and immediate development of Saarinen's proposal.[39] Unlike Kresge or TWA, Dulles' design embraced a form suggested by forces, instead of fighting the forces with form. In describing the process, Saarinen wrote, "we tried to give a completely logical, imaginative, and responsible answer."[40]

The initial developmental drawings showed 16 columns on each side of the open room, spaced 40 feet (12 m) apart, leaning outward with large bases to help resist against the thrust and the roof. Different column heights created a sweeping hanging curved roof that covered the entire 90,000 square foot area (8,361 m²), with a geometry that was defined by the catenary curve of uniformly loaded hanging cables. Because the form expressed an inherent structural logic, very little about the concise and elemental form was required to change for either architectural or structural reasons; the main collaboration efforts were in the translation of this simple design idea into an efficient but expressive structure—specifically the roof and columns.

Instead of trying to cast an inverted concrete shell over a large open area, the design team developed an ingenious method of structuring and constructing a durable and expressive roof using cables and precast panels. The hanging roof cables were suspended between the two upwardly curved, cast-in-place, concrete edge slabs that connected to the columns. At every 10 feet (3 m) of length along the perimeter of the edge slab, two pairs of 1 inch (2.54 cm) steel cables were suspended more than 120 feet (36.5 m) across the span to the other edge beam. Next, 1,800 lightweight precast panels with small metal hooks on the sides were placed one after another upon the cables. The cables were pre-tensioned to the exact funicular geometry, then the roof was weighed down by sand bags until both the cables (in the gap between the panels) and the topping slab on the panels could be encased in a lightweight poured concrete. There was no formwork needed below the roof—ingeniously, the panels were the structure and the formwork for the topping slab.

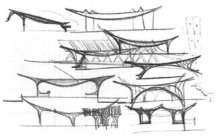

FIGURE 5.7   Dulles Airport Terminal's design process consolidated quickly around a single solution. *Left*: The similarity in Saarinen's initial sketches show his commitment to a large volume with a hanging surface (1958). *Right* and *bottom*: The overall building form and the columns/roof connection shown in the initial models remained consistent with the final building (approx. 1959).

Source: *Top left*: Courtesy of Eero Saarinen Collection, Yale University Archives. *Right* and *bottom*: Images by Balthazar Korab. US Library of Congress, Balthazar Korab Archive.

This extra weight of the panels and topping slab made the entire roof system structurally integral and resistant to wind uplift.[41] Roof drainage and expansion joints were integrated into the system without compromise, and atop the roof surface, an innovative system of layered neoprene membranes was used to preserve the visual clarity of form and material.[42] The precast panels on the interior were sprayed with an insulating acoustical foam to create a visually uniform surface and to improve the building's functional performance (Figure 5.8).

Despite the expressive building form, the repetitive elements and consistent building section only required a conventional number of drawings to document the columns and roof structure. Critical requirements for tensioning the steel cables, setting the precast panels and placing the concrete were thoroughly documented in drawings and specifications. The first line in the specification section "Invitation to Bidders" set the project expectations high by warning that "the work herein proposed for development is highly original in character and will require considerable ingenuity and very skilled workmanship in its construction for obtaining the form and effect contemplated."[43]

FIGURE 5.8 The sculptural form for Dulles expressed resistance to structural forces and effective construction techniques. *Top left*: The columns lean outward to resist the pull of tension forces and thrust. *Top right* and *bottom*: Once the two sides were finished, the hanging cables and precast roof panels spanned across the volume without requiring scaffolding.

Source: Photos by Balthazar Korab. US Library of Congress, Balthazar Korab Archive.

Remarkably, the main roof was constructed without *any* scaffolding below it because of the cable and precast system. Joints between panels were cast in place along with a topping slab to make the roof monolithic. Because the structural elements were regular and repeating, multiple trades worked simultaneously as construction progressed from one end to another. The work flow was so well

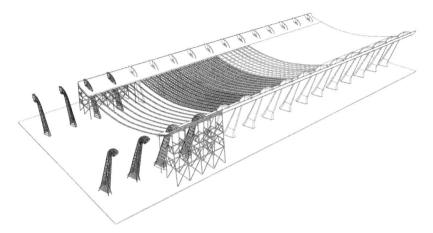

FIGURE 5.9 The form, materials and assembly strategies for Dulles weren't developed in spite of the desired aesthetic and functions of the building, but in correspondence with it. A dramatic evolution after a decade of collaboration.

Source: Illustration by Rob Whitehead and Barrett Peterson.

coordinated that the large steel towers used to cast the edge slabs were simply rolled to their new location and the formwork was re-used. Although the overall project had cost overruns, the construction process was relatively economical thanks to the collective integration of specific strategies for spatial shell construction.

Nearly every complication in design and construction that plagued Kresge was addressed and essentially eliminated at Dulles, less than a decade after Kresge was designed (Figure 5.9). Sadly, Saarinen only lived to see one of the cables hanging across the great space. Months before his death, the last time he visited the construction site, Saarinen said, "I think this terminal building (Dulles) is the best thing I have done . . . Maybe it will even explain what I believe about architecture."[44]

Evolved Influence and Continued Collaborations

These dramatic improvements across the three projects likely results from two major factors. The first is Saarinen's willingness to accept and embrace a greater level of influence and engineering expertise during the early stages of design. Second, an increased level of expertise in concrete shell design and construction developed by the project teams in *both offices* was born from these collaborations and formative failures and setbacks. This collaboration was both unique and instructive. The work advanced much initial disciplinary dialogue about the degree of influence that function, structural performance and constructability should (or shouldn't) have in the derivation of a spatial shell. Additionally, these efforts demonstrated the benefits of a more integrative practice model that transcended the traditional role of design influence and authorship.

After his death, the legacy of Saarinen's practice was continued by his partners in the successor firm, Roche Dinkeloo. They continued collaborating with Ammann & Whitney, and specifically with Boyd Anderson, for decades. As a result of his work with Saarinen, Anderson won an award for "Metropolitan Civil Engineer of the Year" in 1962 and continued to collaborate with many of the nation's leading architects, including Roche, for decades in his role as the head of the Special Structures division of Ammann & Whitney. His efforts developing early engineering methods for shells and tension structures is rightly remembered as being highly influential. After his death in 1995, the ASCE described Anderson's engineering achievements as "legendary" and called him "one of the great structural engineers of our time."[45] Roche summarized the tumultuous but rewarding collaborative efforts concisely: "It was a great moment in modern architecture . . . even if some were appalled by the work, it was incredible."[46]

Notes

1. N. Keith Scott, "M.I.T. Auditorium: An English View," *Journal of the Royal Institute of British Architects* (February 1955): 138.
2. Eero Saarinen, "The Trend of Affairs," *The Technology Review* 57, no. 8 (MIT Press, June 1955): 387–88.
3. Harper's Bazaar Photo proofs, *Eero Saarinen Collection Manuscript and Archives*, Yale University Library, MS 593, Box 168, Folder 503.
4. E. Cohen, N. Dobbs, and W. Combs, "Inspection, Analysis, and Restoration of MIT Kresge Auditorium," in *Rehabilitation, Renovation, and Preservation of Concrete and Masonry Structures*, ACI Special Publication, SP-85 (Detroit: American Concrete Institute, 1985), 92–125.
5. "Saarinen Challenges the Rectangle," *Architectural Forum* (January 1953): 126–33.
6. Eero Saarinen to Aline Louchheim, 1953. Aline and Eero Saarinen Papers, 1906–1977, Archives of American Art, Smithsonian Institution.
7. John Chilton, *Heinz Isler* (London: Thomas Telford, 2000).
8. Anderson Boyd, "Firm Biography," in *Ammann & Whitney Promotional Materials* (New York: Ammann & Whitney, 1958).
9. Kevin Roche interviewed by Rob Whitehead (by telephone), May 27, 2014.
10. Ibid.
11. Rob Whitehead, "Portentous and Predictable: Eero Saarinen, Ammann & Whitney, and the Failures of Kresge Auditorium (1950–1955)," in *Proceedings of the IASS Symposium 2018*, eds. C. Mueller and S. Adriaenssens (Boston: MIT Press, 2018).
12. Stanford Anderson, *Eladio Dieste, Innovation in Structural Art* (New York: Princeton Architectural Press, 2004).
13. Rob Whitehead, "Saarinen's Shells: The Evolution of Engineering Influence," in *Proceedings of IASS-STLE 2014 Symposium*, eds. M. L. R. F. Reyolando and M. O. Ruy (Brasilia, Brazil, 2014). International Association of Spatial and Shell Structures, Headquarters in Madrid, Spain.
14. "Kresge Auditorium Project Specifications," Prepared by Eero Saarinen & Associates, 1952. *Eero Saarinen Collection (MS 593)*, Series IV, Project Records, Manuscripts and Archives, Yale University Library.

15. Pennsylvania State University's Architectural Engineering Lab, *"Case Study #3,"* Historic Preservation of Thin-shell Concrete Structures, 2011.
16. Douglas Bates, "Speech Transcript: Construction of the M.I.T. Auditorium," in *Conference on Thin Concrete Shells* (Boston: MIT Press, June 21–23, 1954), 132–33.
17. "Tripod Built on Tricky Formwork," *Engineering News-Record* (May 27, 1954): 30–32.
18. T. Boothby, M. Parfitt, and C. Roise, "Case Studies in Diagnosis and Repair of Historic Thin-Shell Concrete Structures," *Association for Preservation Technology International* 36, no. 2/3 (2005): 3–11.
19. Charles C. Whitney, "Speech Transcript: Economics," in *Conference on Thin Concrete Shells* (Boston: MIT Press, June 21–23, 1954): 22–24.
20. Eero Saarinen, *General Statement about the Sculptural, Curved Shapes. . .* (undated, 1958–61) Eero Saarinen Collection (MS 593). Manuscripts and Archives, Yale University Library.
21. Eero Saarinen, "Speech Transcript: Function, Structure, and Beauty," *Architectural Association Journal* (July–August 1957): 43.
22. Roche interview, 2014.
23. Christopher Leubkeman, "Form Swallows Function," *Progressive Architecture* (May 1992): 106–8.
24. Eero Saarinen & Associates (no date). "Project Descriptions," *Eero Saarinen Collection (MS 593)*, Series IV, Project Records, Manuscripts and Archives, Yale University Library.
25. Saarinen, *General Statement. . .*
26. Roche interview, 2014.
27. K. Ringli, "Der Ingenieur von Kahn und Saarinen (translated)," *archithese* (May 2011): 54–59.
28. Ibid. and Roche interview, 2014.
29. Richard Knight, *Saarinen's Quest: A Memoir* (Richmond, CA: William Stout Publishers, March 2008).
30. Leubkeman, "Form Swallows Function," 1992 and Roche interview, 2014.
31. Anderson Boyd, A. Tor, and R. W. Yeakel, "Design and Construction of Shell Roof for the New York International Airport TWA Flight Center," in *Proceeding from World Conference on Shell Structures* (Washington, DC: National Academy of Sciences, 1964), 319–28.
32. "Shaping a Two-Acre Sculpture," *Architectural Forum* (August 1960): 119–22.
33. R. Yeakel, "Concrete Shells of TWA's Idlewild Terminal Involves Unique Methods," *Columbia Engineering Quarterly* (January 1962): 16–17, 42, & 47.
34. Ibid.
35. Eero Saarinen to John Peter, December 6, 1960, *Eero Saarinen Collection* (MS 593), Series IV, Project Records, Manuscripts and Archives, Yale University Library.
36. Ed E. Cohen, "Shell Concrete Costs," *Architectural Forum* 103, no. 1 (July 1955): 131, 188.
37. "Four Great Pours," *Architectural Forum* 115, no. 3 (September 1961): 104–15.
38. Eero Saarinen & Associates (no date), "Project Descriptions."
39. Roche interview, 2014.
40. Eero Saarinen to John Peters, 1960.
41. "Four Great Pours," *Architectural Forum.*
42. Lawrence Lessing, "The Diversity of Eero Saarinen," *Architectural Forum* 113, no. 1 (July 1960): 95, 96, 103.
43. "Dulles Airport Terminal Project Specifications," prepared by Eero Saarinen & Associates, 1961, *Eero Saarinen Collection (MS 593),* Series IV, Project Records, Manuscripts and Archives, Yale University Library.
44. Eero Saarinen, as recounted by Aline Saarinen, *Eero Saarinen on his Work: A Selection of Buildings Dating 1947–1964* (New Haven, CT: Yale University Press, 1962).
45. *Transactions of the American Society of Civil Engineers* 160, no. 9 (New York, American Society of Civil Engineers, 1995), 747.
46. Roche interview, 2014.

6

DOING SOMETHING ABOUT THE WEATHER

A Case for Discomfort

Andrew Cruse

"Everybody complains about the weather, but nobody does anything about it," Mark Twain once quipped.[1] Yet today, when human activity is linked directly to a warming planet, this comment no longer seems so true or so amusing. Architecture changes the weather in two principal ways. Carbon emissions from buildings fuel global climate change, while buildings themselves provide stable interior climates designed for human comfort. These two facts work against one another; the carbon emitted to create comfortable indoor environments simultaneously stokes global warming. Architects typically look to energy efficiency measures to reduce this impact, but seldom ask what role comfort itself can play in reducing building energy needs. This is not surprising since the dominant architectural "comfort standard" used in building design suggests that comfort is a problem that has been solved. Such a limited understanding of comfort is ripe for reexamination. A revised idea of comfort that positively connects indoor and outdoor climates can lead to architectural opportunities that pique public consciousness, create thermal delight and reduce building-related carbon emissions.

Ideas of comfort have long been entangled with the weather. In his *On Airs, Waters, and Places* (c. 400 BCE), Greek physician Hippocrates proposed that natural atmospheric systems had a direct influence on human health and comfort. Alexander von Humboldt, and other nineteenth-century natural scientists, documented how the geographic distribution of plant and animal species correlated with different climates. Ideas positively connecting comfort and climate began to shift around the start of the twentieth century, as engineers argued that the artificial climates created by building mechanical systems were superior to outside conditions. This shift was part of a larger change that the sociologist Norbert Elias called the "civilizing process."[2] This process describes the decreasing thresholds of tolerance towards sensuous experiences such as sounds, smells and temperatures that had begun in the Middle Ages. Sensations that were once considered acceptable were later found to be "uncivilized." Two emblematic spaces that help to explain the changing relationship between comfort, indoor climate and outdoor climate are the bathing machine (Figure 6.1) and the psychrometric chamber (Figure 6.2).

FIGURE 6.1   Woman descending from a bathing machine, northern Germany, late nineteenth century.

Source: Wilhelm Dreesen.

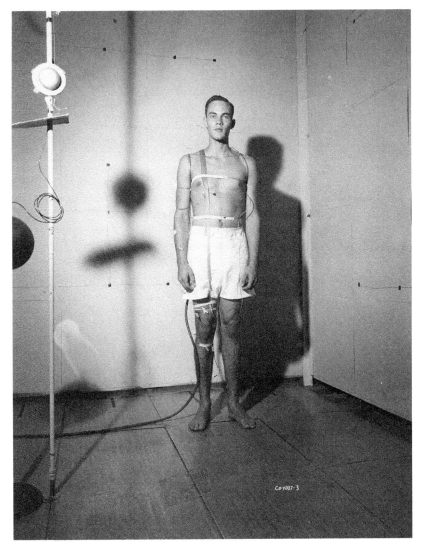

FIGURE 6.2   Man in a psychrometric chamber at Kansas State University's Institute for Environmental Research, mid-twentieth century.

Source: Institute for Environmental Research, Kansas State University.

Sublime Comforts

The bathing machine appeared in the early eighteenth century along the southern coast of England. It developed as part of a "rush to the sea" prompted by the medical community's interest in seawater's salubrious character and the therapeutic fashion for cold water bathing. The thermal and physical shock of water and waves were believed to bring vigor to the bather's body, curing them of gout, jaundice and other glandular problems, as well as treating more existential aliments such as melancholy and "excess spleen." In his *Dissertation on the Uses of Sea-Water in the Diseases of the Glands* (1750), the English physician Richard Russell collected observations from 20 years of experiments to argue that the Divine Creator had made the sea as a common defense against the corruption of the body. His recommended treatments included drinking seawater and, for those who preferred "the rough embraces of Neptune," bathing directly in the sea.[3]

The bathing machine was used to transport paying bathers from the shore to the sea so that they could enter the water without being seen by curious onlookers. There was no standard bathing machine design, but the model built by Benjamin Beale in Margate (Figure 6.3) became popular around the middle of the century. It consisted of a four-wheeled carriage carrying a small wooden cabin with a single door on the gable end. The interior, sometimes illuminated by side windows, had a bench, a mirror and hooks to hang one's clothing. Beale's real innovation was the introduction of a "modesty hood," an awning-like attachment over the door that could be extended to allow bathers to discreetly enter and exit

FIGURE 6.3 Card from John and Mercy Sayer in Margate, former partners of Benjamin Beale, from the 1790s.

Source: Courtesy of Antony Lee.

the water. Sea baths were administered by "dippers," who submerged the bather for the required length and number of times to achieve the prescribed therapeutic benefit. After their bath, bathers reentered the bathing machine to dry off and change into their clothes while the carriage was drawn back up onto the beach. Baths typically took place in the morning, allowing the afternoon to recuperate and pursue leisure activities.

In creating a transition from a fixed, terrestrial world to a mobile, aqueous one, the bathing machine served as an architectural platform for experiencing the sublime as it was described by Edmund Burke.[4] In the eighteenth century, the sea held very different connotations than it does today. Unlike terrestrial landscapes, the sea could not be tamed. It was a remnant of the Biblical Flood, the primitive state of the world that existed before history began. To experience the sea's vastness was to experience the sublime, the combination of awe and terror that emotionally and intellectually moved the subject, compelling them to make sense of their experience.

Sea bathing was part of the larger attention to physical comfort that emerged in eighteenth-century Anglo-American culture.[5] The emphasis on physical comfort—the self-conscious satisfaction with the relationship between one's body and its immediate physical environment—arose in part as an effort to legitimize popular consumption. It came with new types of consumer goods such as mirrors, upholstered furniture and enclosed stoves. The rough comforts of sea bathing were far from these conventional notions of physical comfort. Instead, sea bathing using a bathing machine mixed the comforts of privacy and convenience with the discomforts of exposure and difficulty. As the historian Alain Corbin has discussed, "at the edge of the ocean, modern man comes to discover himself and to experience his own limits in the face of the ocean's emptiness." There "a new world of sensations was growing out of the mixed pain and pleasure of sudden immersion" and "a new way of experiencing one's body was developing, based on rooting out the desires to disturb it."[6] In creating a literal entry into this new world, the bathing machine offers an important example of how the austere comfort of the sublime could pique consciousness, creating literal and conceptual connections between an individual and the larger environment.

Laboratory Comforts

The psychrometric chamber became an essential piece of laboratory equipment for establishing indoor comfort standards during the twentieth century. Also called climate, environmental or air-conditioned chambers, these devices represented the confluence of medical and engineering research concerning the body's relationship to climate. Rapid urbanization in the nineteenth century contributed to poor air quality in buildings, which in turn had a negative effect on occupants' health. Medical theory of the day held that these effects were due to the build-up of carbonic acid or "crowd poison" in the air due to people living and working in close quarters. The solution to this problem was to establish building ventilation standards, specifying a quantity of "fresh air" needed per person in order to replace the "vitiated air" they created by breathing. Physicians generally considered outdoor air brought in through open windows an ideal way to ventilate buildings. Engineers, however argued that artificial ventilation with mechanical equipment was cleaner and healthier than using outside air directly.[7] Such air could be filtered and its temperature controlled, whether it needed to be heated or to be cooled and dehumidified using the newly developed air-conditioning process pioneered by the Carrier Corporation. This confidence in an artificial environment was captured by a marketing booklet published by Carrier (Figure 6.4) entitled *The Story of Manufactured Weather* and authored by the "Mechanical Weather Man," whose valve-like shirt proclaims "every day a good day."[8]

As the use of air conditioning spread, engineers and equipment manufacturers recognized the need for clear comfort guidelines. The fledgling American Society of Heating and Ventilating Engineers (ASHVE) seized upon the opportunity to define comfort as a way to establish their leadership in the field of mechanical engineering. To that end, the Society founded a research laboratory at the Bureau of Mines in Pittsburgh, Pennsylvania, in 1919. There they built a psychrometric chamber to "scientifically" determine the basis for human comfort. The chamber was a windowless room in which temperature and humidity could be precisely set and independently controlled using an air-conditioning system. Researchers had subjects spend time in the chamber under a variety of environmental conditions. Upon exiting, they were asked to rate how comfortable they found those conditions.

Psychrometric chambers like the one built by ASHVE were a necessary piece of equipment in the developing field of industrial hygiene. Industrial hygiene researchers studied how the environmental conditions found in factories and on job sites affected workers' health, from chemical exposure to extremes of temperature and humidity.[9] Psychrometric chambers allowed researchers to predictably recreate industrial environments in a laboratory setting where the response of subjects could be closely studied. ASHVE researchers used the same type of chamber to find ideal interior conditions for white collar environments.

The physiological premise for ASHVE's research was that comfort results from the heat balance between a body and its external environment. That is, when the heat produced by metabolism was balanced with a room's air temperature and humidity, people felt comfortable. Researchers hypothesized that there must be different combinations of temperature and humidity that produced equally comfortable conditions. By 1924, researchers at the Bureau published a "Comfort Zone" chart based on data obtained using their psychrometric chamber. The chart showed a limited band of temperature and humidity in which half of the occupants felt comfortable, as well as a comfort line at which 97 percent of test subjects found the conditions comfortable (Figure 6.5).[10]

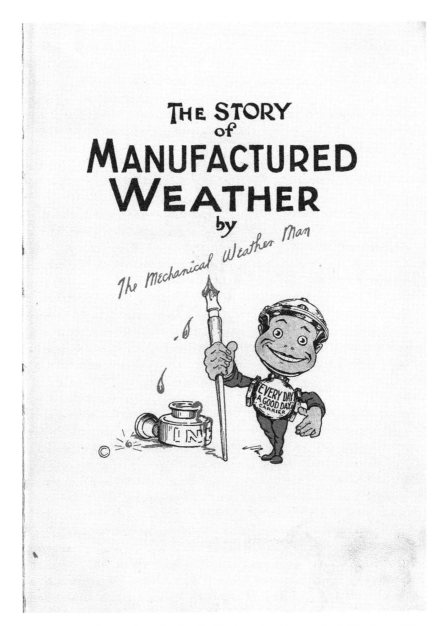

FIGURE 6.4 Title page from the Carrier Engineering Corporation's *The Story of Manufactured Weather by the Mechanical Weather Man*, 1919.

Source: Author.

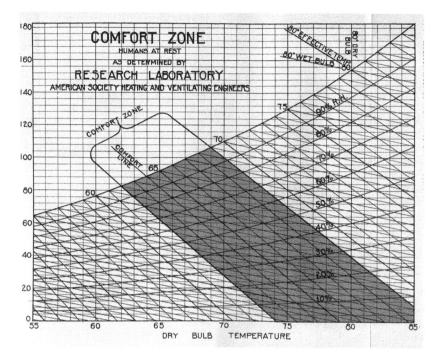

FIGURE 6.5 Comfort Zone: Humans at Rest as Determined by Research Laboratory American Society Heating and Ventilating Engineers.

Source: Journal of the American Society of Heating and Ventilating Engineers, March 1924, n.p.

Buildings soon began to resemble psychrometric chambers as controlled interior climates were separated from exterior ones. In many cases, this meant building occupants were not allowed to open windows. As one commentator succinctly put it, when a building has operable windows and no air conditioning, employees tend to blame nature for their discomforts. But when a building has air conditioning, they blame the building's management.[11] Tenants in the Milam Building (San Antonio, Texas, 1928), the first air-conditioned office building, were prohibited by their lease agreements from opening the windows in their offices because it made managing the air-conditioning system too difficult. This meant that room-scaled individual comfort control was surrendered to the building engineer, who was responsible for providing comfort conditions for everyone. In more extreme examples, windows were eliminated from buildings. The Modern Office Building (1935), built for the Hershey Chocolate Corporation in Hershey Pennsylvania, was touted for its air-conditioned comfort and energy efficiency (Figure 6.6). Outdoor weather information was communicated to building occupants through a series of colored lights below the wall clocks.

FIGURE 6.6 Postcard showing the Modern Office Building of the Hershey Chocolate Corporation, Hershey Pennsylvania. A 2015 renovation added 114 windows to the building.

Source: Author.

Despite continued comfort research through the 1950s, ASHVE's conclusions about ideal comfort conditions became less certain. (ASHVE merged with another organization in 1959 to form ASHRAE, the American Society of Heating, Refrigeration and Air Conditioning Engineers.) ASHRAE's 1961 comfort chart admitted "there is no precise physiologic observation by which comfort can be evaluated."[12] This uncertainty led to a growing frustration among members, who wanted clear direction upon which to base their designs. Recognizing that the comfort zone was losing the authority it once held for the mechanical engineering profession, ASHRAE began planning a comprehensive reevaluation of comfort.

In this context of uncertainty, Danish researcher P. O. Fanger proposed his Predicted Mean Vote (PMV) thermal comfort model. Rather than the graphic method of the comfort chart that addressed the two variables of temperature and humidity, Fanger developed a comfort equation based in part on psychrometric chamber research. This included six variables, four environmental parameters—dry bulb temperature, relative humidity, mean radiant temperature and air velocity—as well as two individual ones—activity level and amount of clothing. Fanger claimed that research using psychrometric chambers indicated no national, geographic or seasonal variation in comfort condition. In fact, he asserted that any deviation from a near steady state interior climate would only result in a greater number of people being dissatisfied.[13]

Fanger's work achieved the initial goal set for the Research Bureau of establishing a practical comfort standard that was grounded in laboratory research and useful to practicing engineers. In his book *Thermal Comfort*, Fanger wrote, "thermal comfort is the 'product' which is produced and sold to the customer by the heating and air conditioning industry. It is therefore obvious that quantitatively expressed comfort conditions are of great importance."[14] Fanger's PMV model was widely accepted by the engineering profession and the equipment industry, and today it underlies two of the most widely used engineering comfort standards, the ASHRAE 55 "Thermal Environmental Conditions for Human Occupancy" and ISO 7730 "Moderate Thermal Environments."

The twentieth-century development of architectural comfort embodied in standard engineering practice, mechanical equipment and the spaces that they condition has reduced the contemporary understanding of comfort to a neutral heat balance maintained in part by the clear separation between indoor and outdoor climates.

Adaptive Comforts

The bathing machine and the psychrometric chamber provide opposite conditions for comfort. The bathing machine creates a platform for the austere comfort of the sublime while the psychrometric chamber provides an invisible comfort of thermal neutrality. By mixing interior and exterior climates, the adaptive comfort model represents a loose synthesis of these two.

Beginning in the 1960s, building professionals sought novel ways to expand the homeostatic indoor climate prescribed by the PMV model by bringing dynamic elements of the outdoor climate to the interior. This interest was driven in part by the heightened environmental awareness of the 1960s and the increasing energy costs due to the oil embargo of the early 1970s. The approach developed from field studies in which researchers asked occupants of existing buildings to rate how comfortable they found the interior climates. Unlike the laboratory-based psychrometric chamber work, these field studies reflected real-world conditions, with a greater diversity of populations, clothing levels and metabolic rates. Generally, these studies demonstrated that occupants considered themselves comfortable over a wider range of temperatures in naturally ventilated buildings than were predicted by the PMV comfort model. This research led to the development of the Adaptive Thermal Comfort (ATC) model.[15]

ATC researchers proposed that individual comfort resulted from feedback loops between the environment and its occupants. In this view, occupants were not passive recipients of comfort conditions, as held by the PMV model, but were actively engaged in creating their own comfort. Researchers identified behavioral, physiological and psychological adaptive mechanisms as influencing occupant comfort. Behavioral adjustments include changes people can make to their immediate environment, such as taking off a sweater or opening a window.

Physiological adjustments are changes in how the body reacts to the thermal environment. They range from short-term acclimatization that happens in a matter of weeks or months to long-term genetic changes that happen on an evolutionary time scale. Psychological changes are based on past experience, such as seasonal or cultural expectations, and result in changing reactions to thermal information.

In addition, ATC researchers showed that the thermal monotony associated with the PMV model was energy intensive. By allowing temperatures to float over a wider range that more closely correlated with the weather outside, they demonstrated that buildings used less energy. Later research qualified that, in commercial buildings, the heating and cooling energy use changes roughly 7 percent for each 1.8°F shift out of the thermally neutral zone.[16]

Recent medical thinking also calls into question the desirability of thermally neutral interior environments. Following advances of the past 50 years, lower human mortality from deadly diseases has been replaced by higher morbidity from diseases of excess such as Type 2 diabetes and obesity. Today's dominant notion of thermal comfort contributes to this morbidity by minimizing the amount of energy needed by the body to maintain heat balance with the surrounding environment. In a situation with limited resources, the body has evolved to favor environments requiring less energy. Yet today's thermally neutral interiors, where the body achieves heat balance at little energetic cost, have led to mismatches between comfort and well-being, in that remaining indoors and largely sedentary is a leading cause of disease and disability.[17] Biologists and doctors remind us that we often mistake comfort for well-being.[18] From recommendations to increase physical activity and to perform challenging mental tasks, to enthusiasm for extreme feats of endurance and exposure, today's culture of wellness is fostering a growing awareness that some degree of stress is an important part of well-being.

Although the ATC model established a connection between exterior conditions and interior comfort, it does not explain why the same environmental conditions found unacceptable in air-conditioned buildings are found acceptable in naturally ventilated ones. To answer this question, researchers have turned to the idea of alliesthesia. Initially proposed by the French physiologist Michel Cabanac, alliesthesia describes how a stimulus can be regarded as either painful or pleasant depending on the subject's internal state.[19] In psychological terms, alliesthesia describes the mechanism of perception—the interpretation of a stimulus—rather than sensation—the recognition of a stimulus. Take for example two subjects, one with a slightly elevated body temperature, say after exercise, and one with a slightly lowered body temperature, such as coming inside on a winter day. If both subjects were to put their hands into water at 50° F, both would judge the water to be cold based on the process of sensation. However, the subject with the elevated body temperature would find the perception of coolness pleasant due to their elevated body temperature, while the subject with the lower body temperature would find it unpleasant due to their lowered body temperature. Based on this observation, Cabanac argued that thermal pleasure was only found in transient states where the temperature exposure moves a subject towards a normal body temperature.

Based on this observation, adaptive comfort researchers argue that thermal pleasure is unachievable in a neutral state. That is, the thermally neutral conditions prescribed by the PMV model are incapable of providing pleasure since they are based on a static heat balance between environment and body. Alliesthesia shows that there needs to be thermal differences, however small, between the subject and the environment to create pleasure.[20] According to the PMV model, such thermal differences are referred to as "localized discomfort" since they shift the body from a neutral state. Air movement is unfavorably labeled as "draft." Asymmetric thermal conditions and temperature drifts are seen as disturbing the body's uniform heat balance. Alliesthesia suggests how designers can shift consideration of such discomforts to see them as pleasurable.

Discomfort and Delight

Discomfort disturbs the calm surface of a thermally neutral world to make perceptible the edges of comfort, bringing it to the attention of building users. In this sense, discomfort is not comfort's opposite, but its complement.[21] The bathing machine offers an extreme version of such differences related to the sublime. Using vocabulary similar to Cabanac's, Burke contrasted the sublime's affective experiences with the banality of indifference. The sublime's pain at a distance rendered it pleasurable, just as in alliesthesia the difference between internal and external states led to thermal pleasure.[22]

Discomfort can inform design to heighten awareness and expand what we expect of the interior climate. The installation Cloudscapes, by architect Tetsuo Kondo and Transsolar at the 2010 Venice Biennale, created a cloud which magically floated in the vast brick hall of the Arsenale building (Figure 6.7). Visitors experienced this opaque yet ephemeral vapor by walking along a 43-meter ramp that took them from the ground, up through the cloud to an open space under the ceiling, and back down again. Although the process of forming and maintaining the cloud required sophisticated mechanical equipment, visitors only perceived the resulting environment.[23] Moving through Cloudscapes' thermal strata—from the cool, dry air closer to the ground, to the warm, humid vapor of the cloud and to the dryer, warmer air just under the roof—visitors become aware of their own thermal experiences. Cloudscapes, like the bathing machine, opened up a new realm of experience that heightened one's awareness of one's own body and of the environment surrounding it. The exhibition, like climate change, disturbs the expected equilibrium between body and built environment in a visceral way that piques consciousness through discomfort.

Expanding established notions of thermal comfort derived from the psychrometric chamber moves our understanding of comfort in an "uncivilizing"

FIGURE 6.7   Cloudscapes installation in the Arsenale at the 2010 Venice Architecture
Biennale by Tetsuo Kondo Architects and Transsolar.

Source: Courtesy of Tetsuo Kondo.

direction towards discomfort—those very conditions Elias saw as being eliminated from the modern world. Conditions that are routinely excluded from
interior climates designed for thermal neutrality should instead be welcomed as
sources of well-being and pleasure. Like the bathing machine, these environments
disquiet our thermal expectations and pique our consciousness to focus on the
potential of discomfort.

Consciously addressing the construction of comfort allows architects to
do something about the weather in deliberate and productive ways. Ideas of
adaptive thermal comfort, alliesthesia and discomfort can redefine contemporary understandings of comfort. They demonstrate how people can adapt to
more varied indoor climates. They show how such variety can provide thermal pleasure in ways not found in a thermally neutral world. And they can
reduce building energy usage by expanding interior climates to more closely
follow the exterior weather. These ideas highlight how comfort, like climate,
is not a stable index of energetic balance but rather a condition of flux on
which human activity and decisions have a direct impact. The task then falls to
architects to translate such ideas into built form that explores comfort's poetic
potential while provoking public debate about positively changing the climate
around us.

Notes

1. Although this quotation is generally attributed to Twain, its true author remains a debatable subject. See *Respectfully Quoted* (Mineola, NY: Dover Publications, 2010).
2. Norbert Elias, *The Civilizing Process* (New York: Wiley, 2000).
3. Edmund W. Gilbert, *Old Ocean's Bauble* (London: Methuen, 1954), 72.
4. Edmund Burke, *A Philosophical Enquiry Into the Sublime and the Beautiful* (New York: Oxford University Press, 1990).
5. John E. Crowley, *The Invention of Comfort: Sensibilities & Design in Early Modern Britain & Early America* (Baltimore: The Johns Hopkins University Press, 2001).
6. Alain Corbin, *The Lure of the Sea: The Discovery of the Seaside, 1750–1840*, trans. Jocelyn Phelps (New York: Penguin Books, 1994), 95.
7. Gail Cooper, *Air-Conditioning America: Engineers and the Controlled Environment, 1900–1960* (Baltimore: Johns Hopkins University Press, 1998), 51–79.
8. Carrier Engineering Corporation, *The Story of Manufactured Weather by the Mechanical Weather Man* (New York: Carrier Engineering Corporation, 1919).
9. Christopher Sellers, *Hazards of the Job: From Industrial Disease to Environmental Health Science* (Chapel Hill: University of North Carolina Press, 1977), 141–86.
10. ASHVE, "Celebrate Thirty Years of Progress," *Journal of the American Society of Heating and Ventilating Engineers* 30, no. 3 (1924): n.p.
11. Charles S. Leopold, "Conditions for Comfort," *ASHVE Transactions* 53 (1947): 295–306.
12. Ralph Nevins, "Psychrometrics and Modern Comfort," *ASHRAE Transactions* 67 (1961): 609–21.
13. Povl Ole Fanger, *Thermal Comfort: Analysis and Applications in Environmental Engineering* (New York: McGraw-Hill Book Company, 1972), 85–95.
14. Ibid., 15.
15. Richard de Dear and Gail Brager, "Towards an Adaptive Model of Thermal Comfort and Preference," *ASHRAE Transactions* 104, no. 1 (1998): 145–67.
16. Edward Arens, Michael A. Humphreys, Richard de Dear and Hui Zhang, "Are 'Class A' Temperature Requirements Realistic or Desirable?" *Building and Environment* 45, no. 1 (2010): 4–10.
17. "Physical Inactivity a Leading Cause of Disease and Disability, Warns WHO," accessed October 2019, www.who.int/mediacentre/news/releases/release23/en/.
18. Daniel Lieberman, *The Story of the Human Body* (New York: Penguin Books, 2013), 318–46.
19. Michel Cabanac, "Physiological Role of Pleasure," *Science* 173, no. 4002 (September 17, 1971): 1103–7.
20. Richard de Dear, "Revisiting an Old Hypothesis of Human Thermal Perception: Alliesthesia," *Building Research & Information* 39, no. 2 (2011): 108–17.
21. For two recent considerations of discomfort, see David Ellison and Andrew Leach, eds., *On Discomfort: Moments in a Modern History of Architectural Comfort* (New York: Routledge, 2017) and Jacques Pezeu-Massabuau, *A Philosophy of Discomfort*, trans. Vivian Sky Rehberg (London: Reaktion Books, 2012).
22. Burke, *A Philosophical Enquiry Into the Sublime and the Beautiful*, 28–29.
23. Nadir Abdessemed and Matthias Schuler, *Transsolar KlimaEngineering Tetsuo Kondo Architects Cloudscapes* (Ostfildern: Dr. Cantz'sche Druckerei, 2010).

PART 2

Assembling Constructions

7

RESPONSIVE MODERNISM

Louis Kahn's Weiss Residence Enclosure

Clifton Fordham

Louis Kahn's three-month trip to Europe in December of 1950 as a Fellow at the American Academy in Rome represents a clear and convenient break in his career as an architect. During the trip, Kahn was able to connect directly with ancient sites, an experience that solidified his appreciation for the monumental and time-less. Kahn was 50 years old when he embarked for Rome and up until that point had enjoyed a moderately successful career as a practitioner, with a small number of built works that would hardly anticipate the notoriety gained later in his career and after his death in 1972. Buildings that Kahn designed after the Rome trip are now considered masterpieces by a genius who blossomed later in life.

The paucity of contemporary accounts of Kahn's earlier projects, before his fellowship, have cast a shadow on the quality of his earlier work. Among the works that deserve attention is the Weiss Residence (Figure 7.1), designed and built from 1947 to 1950, before Kahn's Rome trip, and finished shortly after he returned. Located on the outskirts of Philadelphia, it embodies characteristics of the monumental *and* technical inventiveness with building assemblies, material, systems integration and solar control. The formal and technical characteristics of the Weiss Residence, particularly its integrated solar control, counter the notion of the Rome trip as a clean break and the insignificance of Kahn's work before it.

The Weiss Residence offered a promising alternative to shortcomings inherent in functional modernism and incorporated expressive novel window assemblies that mediated light and views. Kahn's design acknowledged human needs and demonstrated a responsiveness to specific environmental conditions in a more overt manner than in later projects. He accomplished these goals and breached the norms of International Style material use and detailing, reflecting theories expressed in his essay "Monumentality," published six years prior to the Rome trip in the 1944 book *New Architecture and City Planning, A Symposium.*[1]

FIGURE 7.1 Photo of the Weiss Residence, southern facade.

Source: Louis I. Kahn Collection, University of Pennsylvania Architectural and Pennsylvania Historical and Museum Commission.

Accent

A series of events that charged Kahn's meteoric rise began in the autumn of 1947. Kahn started a position as a visiting design critic at Yale University when the previous candidate, Oscar Niemeyer, was not able to fulfill the position after he was denied entry to the country. The Architecture Department at Yale was located in the School of Fine Arts, and Kahn's weekly trips from Philadelphia to New Haven led to extended conversations with leading postwar artists, including Joseph Albers, that contributed to a shift in his ideas of architecture toward the material.[2] In the fall of 1950, George Howe was appointed the Architecture Department chairman. Kahn had collaborated with Howe from 1938 to 1942, and Howe was impressed enough with the younger Kahn to actively support the fellowship in Rome.[3]

Kahn's association with Yale and Howe also led to the opportunity to design an addition for Yale University Art Gallery, a commission that arose when the previous architect, Philip Goodwin, was dismissed.[4] The project is largely understood to represent the beginning of the mature period of Kahn's career. It helped launch him toward fame outside of academia and the community of American architects working within the paradigm of postwar International Style architecture with which he had been previously associated.

Reasons for highlighting the Rome juncture include several distinct differences between Kahn's pre-Rome career and the Yale commission. For example, Kahn had not completed any institutional buildings on his own, although construction had commenced for a pair of Philadelphia Psychological Hospital additions for which he was responsible. Kahn's pre-Rome buildings, including the hospital additions, had more idiosyncratic massing than the Yale Gallery, and his earlier work eludes categorization as monumental. Another factor, not typically cited, is that overall building forms and the underlying order of Kahn's earlier designs are more difficult to comprehend in a single photograph than his later, more monumental work, since the irregular massing did not contribute to a quick reading of the overall form. In contrast, Kahn's monumental buildings utilized fundamental massing shapes in plan, elevation and section.

Kahn's post-Rome work is also associated with technological innovativeness, primarily in structural concrete, which contributes significantly to each building's expressiveness. The exposed triangular waffle floor slab in the Yale Gallery and the exposed concrete cantilevers of the Richard's Medical Laboratory (1957–60) at the University of Pennsylvania are two prominent examples. The structures of these buildings pushed convention, requiring significant engineering expertise that wealthy institutional clients could support. Kahn also developed a clear relationship between the primary structural systems that represented his concept of "order" and contemporary mechanical, electrical and plumbing systems that in his opinion undermined order and architectural space. By making environmental systems inconspicuous and subservient to architectural order and space, Kahn's buildings gained a fundamental reading when viewed as a whole.[5] Up close, the less permanent, imperfect materials and handwork of the building surfaces created a human scale syntax that most postwar buildings lack.[6]

Kahn's formal monumentalism and material use offered an alternative to the smooth surfaces of mid-century modernism that reflected the technology of industrialization, often at the expense of local materials, traditions and environments. His monumentalism also addressed a human need for meaning, symbolism and a connection to something larger than the immediate. However, Kahn's post-Rome approach to design does not provide a clearly replicable template for buildings that do not warrant the symbolic weight of public institutions. Even to this end, most contemporary practitioners of architecture have failed to even remotely emulate Kahn's approach, resulting in iconic forms without nuanced humanistic details and connection to local conditions. Kahn's earlier, less symbolic work also addressed much of the criticism that mid-century modernism was inhuman and banal. His background and experiments prior to Rome reflected a gradual transition as opposed to an abrupt one. Kahn's work during this period offers a unique form of modernism that tentatively fits with its stylistic framework and is applicable to complex and asymmetrical design problems. It is also one that is responsive to human needs, including physical and psychological comfort, as well as a sense of rootedness.

Modernism and Designing With the Sun

Louis Kahn's academic training at the University of Pennsylvania, where he studied under Paul Cret, was in the Beaux-Arts tradition. However, his career as a practitioner was shaped by his association with contemporaries who practiced an American derivative of early twentieth-century modernism. Like the European modernism exemplified by Le Corbusier, American avant-garde modernists, many of whom were European émigrés, were drawn to associations between machines and buildings, and against traditional ornament. As a result, the shape of functional architecture reflected the particulars of specific programs, in lieu of specific typologies.

Architects such as Walter Gropius, who settled in New England, tailored their architecture to accommodate local building technologies, like platform framing and materials, including clapboard siding. However, lack of ornament further confused the comprehension of program type and place, contributing to a crisis in architecture manifested after the war. For these American architects, technology and tradition could be reconciled. Gropius's own house (1938) in Lincoln, Massachusetts, had an abstract plan and rectilinear volume, with a flat roof (Figure 7.2). Its siding, echoed on the interior, was drawn from New England vernacular structures. The glass expanse of the southern facade captured sunlight, and Gropius incorporated elements specific to modern architecture for solar

FIGURE 7.2   Gropius House, Lincoln, Massachusetts, Walter Gropius Architect.

Source: Photograph by author.

control guided by science. The features included screened exterior partitions and overhangs, strategically placed to provide shade at appropriate times of the day.

Kahn's architectural legacy is very closely linked to the dynamics of natural light. He verbalized this stance in his address to the Otterlo Congress in 1959, where he stated,

> no space is really an architectural space unless it has natural light. Artificial light does not light a space in architecture because it must have feeling in it from the time of day and season of the year—the nuances of this are incompatible with the single moment of an electric bulb.[7]

Kahn is not remembered as an advocate of solar architecture, but his activities during World War II and immediately after the war placed him in an architectural community that actively sought solar design solutions. Solar architecture permeated a context in which artificially conditioned spaces before the war were seldom air-conditioned mechanically, and indoor and outdoor environments were more fluid.[8]

The first examples of Kahn designing in response to the dynamics of natural light are evident in works that resulted from his collaborations with George Howe, co-designer of the PSFS Building, and Oscar Storonov, a German émigré and scholar of Le Corbusier. Howe's modernist tendencies developed late in his career and, like Gropius, incorporated local building traditions. Kahn's professional partnership with Howe in 1941 was brief but resulted in his first celebrated built work of modern architecture, the Carver Housing complex in rural Pennsylvania, finished in 1943. Many of the buildings in the complex have deep overhangs, covered entries and trellised screen walls with vertical louvers positioned to shade exterior areas against the sun when low it is on the horizon. Similarly, Kahn's unbuilt scheme for Unity House (1945–47), designed when Kahn was a partner with Storonov, utilized similar elements. Kahn's visionary projects included the 1944 Parasol House study for the Knoll Company, which incorporated broad overhangs and operable windows with sliding screens—an experiment that was continued in the Weiss House.

Kahn's use of solar shading devices in the Weiss Residence was anticipated by his entry for the Pennsylvania Solar House project, designed while he was in partnership with Storonov. In 1946, the Libby Owens-Ford company sponsored the project, which included 48 participants from each state, as a method of promoting interest in their proprietary insulated glass product, branded as Thermopane. George Howe was a judge for the project and invited the Kahn & Storonov firm to participate. Kahn and Anne Tyng, a young architect who joined the practice immediately before the commission, were the primary designers, with Tyng's imprint solidly on the project.

The Solar House has a tight service core and a trapezoidal symmetrical plan that was likely brought to the project by Tyng, who studied at Harvard, where

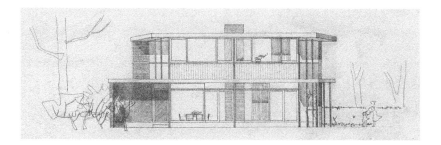

FIGURE 7.3 Pennsylvania Solar House south elevation, Kahn & Storonov Architects.

Source: Louis I. Kahn Collection, University of Pennsylvania Architectural and Pennsylvania Historical and Museum Commission.

she was influenced by Gropius's disciple Marcel Breuer. It includes an entry at the back of the house with few openings, in this case due north, facing the public road. The south side of the house is almost completely clad in glass, while the eastern and western facades are roughly 50 percent glazed (Figure 7.3). Among the more astute techniques is the use of horizontal shading devices, plus deep roof eaves on all faces of the building with the exception of the north face. This shift in facade treatment reflected efficiencies and honesty, communicating the purposefulness of individual elements. The western elevation incorporates two techniques that reflect sensitivity to the western sun, which is difficult to shade due to its position low on the horizon. A sun room is situated off the living room and has a vertical trellis at the west side.[9] The remaining windows were fitted with exterior vertical accordion panels that can be closed to provide complete solar protection and privacy.

The collaboration between Kahn and Tyng merged Kahn's intuition and openness to new technologies with a rational approach to design brought by Tyng. To support their design, they presented drawings (Figure 7.4) that demonstrated the impact of the house's sun shading by illustrating how daylight entered the interior at different times of the day and year. Such a scientific method to assess the impact of sun position was uncharacteristic of Kahn, and is evidence of his tendency to learn from those around him.

The Weiss Residence

Storonov and Kahn's partnership was dissolved in March 1947, sparked by a dispute over design credit for the Solar House.[10] The Weiss Residence, which represented a fresh start for Kahn, is the first residential building designed and constructed after he formed his solo practice.[11] The clients, Morton and Lenore Weiss, were likely connected to Kahn through their synagogue when Kahn and Oscar Storonov were considered for the design of a new worship space and community center. Morton inherited and operated a men's clothing store on Main

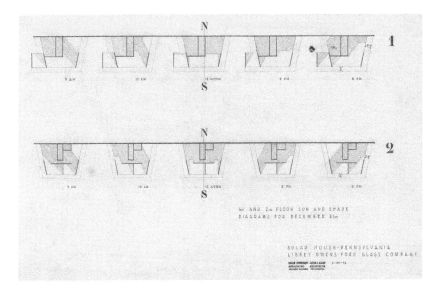

FIGURE 7.4   Pennsylvania Solar House diagrams, Kahn & Storonov Architects.

Source: Louis I. Kahn Collection, University of Pennsylvania Architectural and Pennsylvania Historical and Museum Commission.

Street in Norristown, Pennsylvania, a town approximately 6 miles northwest of Philadelphia. The childless couple selected a large elevated property 4 miles from Norristown, adjacent to a large thoroughfare.[12] The gentle slope of the site upward from the road and the neighboring farms contributed to a sense of openness, offering ample sunlight and views. A design contract was recorded on October 24, 1947, and Kahn began working on the project the next month, with the bulk of the design work occurring in the first half of 1948.[13]

Assuring privacy, Kahn placed the house at the northeastern corner of the site over 450 feet from, and at the highest point (30 feet) above, the main thoroughfare. A driveway was situated parallel to the eastern edge of the site so that the approach passes the house to the east heading northwest, wrapping the house and terminating to the north of the living quarters in front of a separate garage that was connected by a covered walkway.[14] The approach reinforced two important aspects of the project—an uninterrupted view toward the south across an open field down toward the thoroughfare, and relative privacy at the north elevation of the main house, which was close to the adjacent properties. Kahn placed the dining room, kitchen, entry and auxiliary bedrooms on the north side (Figure 7.5).[15]

Working under the rubric of functional modernism, Kahn shaped the building plan to correspond to, and anticipate, human activity. The planning strategy for the main house distinguished related functional spaces in different proportioned rectangular volumes. In plan, the two primary rectangles can be perceived

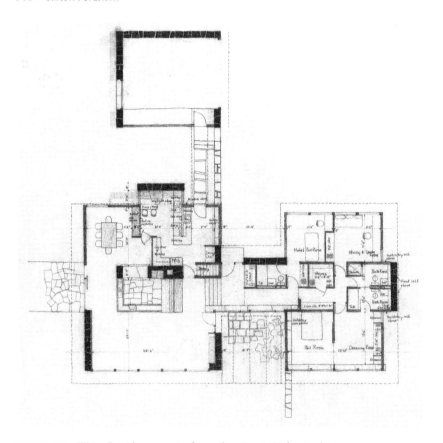

FIGURE 7.5 Weiss Residence main floor plan, Louis Kahn Architect.

Source: Louis I. Kahn Collection, University of Pennsylvania Architectural and Pennsylvania Historical and Museum Commission.

as a single volume created by carving recesses out of a singular mass, or as two masses jointed by a connecting spine. A recessed entry vestibule and bathroom was situated within the spine. The southern recess bounded by the two rectangular masses and the connector formed an informal sitting court and transition space to the southern field. A rectangular zone in plan that includes the connector mechanical room, bathrooms, closets and circulation penetrates into the living volume and the bedroom volume in plan connecting the sleeping wing to the living wing. The service spine is recognizable at the east end of the building, where the mass enclosing the bathroom protrudes beyond the bedroom walls to the edge of the roof, and slightly beyond. The sequestering of support spaces in the center zone, and the distinction between the garage, and main part of the house, anticipated Kahn's use of served and service zones, which became a defining aspect of his work.[16]

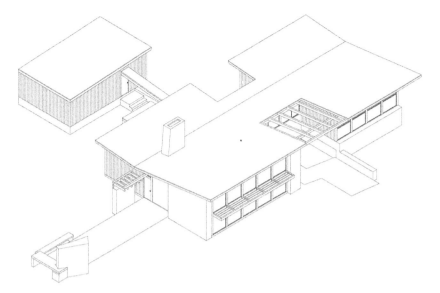

FIGURE 7.6   Axonometric drawing of the Weiss Residence.

Source: Drawing by Clifton Fordham and Peehu Sinha.

Kahn also was able to unify the form by extending a single roof, in this case a long reverse pitched roof, over the main house (Figure 7.6). A reverse pitch roof, commonly known as a "butterfly roof," became a regular component of Marcel Breuer's house designs starting in the mid-1940s. Use of this type of roof in the Weiss Residence was a low-tech means of accomplishing a few important goals. First, the overall roof form compensated for the plan's eccentricities, including recesses and different window sizes. Second, it reinforced the central axis of the building, which is important for navigating the interior. Third, it contributed to a functional hierarchy in the building; the interior spaces are intimate, and the spaces near the high point of the roof, at the perimeter, are located so that they have a strong connection to natural light and openness of the natural world.

Tuned Elevations

The major building elevations reflect a sensitivity to site, interior function and relationships between inside and outside, a task that is more difficult with uniform glazing and materials strategies. This responsiveness is evident in asymmetrical glazing treatment that reflects the interior functions in a more direct way than Kahn's later, more monumental works. For example, large expanses of glazing were assigned to living, dining and sleeping spaces and favor views into the landscape, away from the driveway. Windows in these spaces are constructed of wood, reinforcing a hierarchy of programmatic importance. Glazing in service areas, including the kitchen, laundry and bathrooms, is framed in steel.

Many of the steel windows are transom awning windows, including the bathrooms and kitchen, which strategically allow for cross breezes while supporting privacy. A transom hopper was placed above the main entry for cross breezes and the same type of window is used in the kitchen. For the dining room, Kahn located a vertical fixed window at the corner of the room, adjacent to a custom-designed table that extends to the wall. The most prominent part of the elevations, apart from the roof and associated trellises, are large sections of adjacent window units situated between robust trim that conceals structure. They appear in three locations, at the living room, master bedroom and guest bedrooms. The windows at the southern side of the living room are taller and include a louvered overhang at the middle that corresponds in depth to the eave extension, and divides the window system into two equal sections (Figure 7.1). Intelligent placement of shading and operable venting was critical for comfort considering that mechanical air conditioning was not introduced into the project. In the late fifties, air conditioning was not typical in residential construction, but the technology was feasible for individuals of the Weisses' means.

Materials and Details

Throughout his career, at the scale of the building, Kahn would continue to use non-representational abstract forms that were central for modernism, but when rendered smoothly seemed ungrounded in the passage of time. He compensated by deftly introducing heavy, rough materials into an otherwise abstract language, permitting him to imbue his designs with a sense of weight and permanence. With the Weiss Residence, Kahn's use of stone fulfilled this goal. However, heavy materials that appeared perfect would betray the notion of time. Kahn was able to rectify this by allowing the human hand, and thus imperfection, into his buildings by not resolving every juncture perfectly.[17] The use of rough materials allowed the surfaces of his buildings to "speak" despite not utilizing traditional ornament.

With the Weiss House, Kahn began a quest to establish meaning in architectural details that would remain a key component of his work for the remainder of his life. He also began to contrast heavy, more permanent materials with less permanent infill, which in this case are wood cladding and window trim. The primary exterior cladding material is comprised of 12-inch-wide vertically oriented tongue-and-groove cypress boards. On the western elevation, eastern elevation and portions of the southern elevation, 12-inch-thick stone rubble from a quarry adjacent to the site was used. The project represents Kahn's first foray into a large reveal that he called a "shadow joint."[18] Mortar joints between the stone pieces are deeply recessed to accentuate the profile and individual character of each piece of the stone. All of the stone touches the ground, forming a base below windows and wood siding. The effect is to ground the building in the local condition, a technique that was exceptional for International Style buildings that emphasized lightness.

The wood-clad main windows contribute the weightiness of the facade. In contrast, the windows at the dining room, kitchen, entry vestibule and eastern-most bathrooms are thin profile steel framed, reflecting the more functional aspect of their roles and orientation toward the rear of the house. The deeper wood window frames and relationship of the glazing to the trim face also contributed to a greater reading of depth in the facades as compared to earlier work.

Kahn was careful not to suggest that infill elements were structural elements supporting the roof system. Evidence of this is available at the western masonry wall, where in a preliminary scheme the stone continued all the way to the roof, a situation that would have necessitated cutting it to correspond to the flat surface of the roof soffit. In the final scheme, Kahn lowered the non-bearing wall down to the bottom of the rafter, and capped the stone wall with slate, exposing a wood fascia outside of the wood rafter at the end of the building. A result of holding the stone wall away from the wood rafter is that Kahn did not have to resolve the rough nature of the stone with the underside of the roof projection by finishing the stone so that its edge followed the roof line. The large "shadow joint" makes up the difference.

Integrated Window System

The most innovative part of the house is the functional window system that Kahn introduced at the bedrooms and the living room. Privacy, which is warranted in bedrooms, is a condition that was discounted in many modern masterworks, including Mies van der Rohe's design for the Farnsworth House, where all of the exterior windows span floor to ceiling. Uniform windows, often inoperable, are conducive to the dematerialization of elevations in which specific elements such as frames and hardware do not attract attention. Such effects, privileging exterior reads of buildings, are more tolerable in institutional settings where the image of a building is an important communication reality. In residential structures, living patterns often demand a response other than complete transparency or demate-rialization. Regularity of apertures, and lack of window operation, is contrary to human needs and undermines environment benefits, practical and sensory, that ventilation and natural breezes provide.

At the bedrooms and living room, Kahn introduced robustly framed oversized double-hung windows with additional functionality relating to the importance of the spaces. These windows were inserted between structural framing spaced on a 5-foot 4-inch wide grid.[19] A grand height of the windows (13 feet, 4 inches high) was possible because of the sloped butterfly roof. The windows at the bedrooms rest on a 30-inch high base and extend to the underside of the roof framing. At the living room, the window system (Figure 7.7) extends from the floor slab to the underside of the roof framing and includes an integrated exterior shading

FIGURE 7.7   Interior photo of the Weiss Residence living room.

Source: Louis I. Kahn Collection, University of Pennsylvania Architectural and Pennsylvania Historical and Museum Commission.

device and corresponding light shelf at the inside. This leaves a 6-foot, 4-inch clear space below the shelf.

Together, the bottom and top window sashes in each bay can slide from the top to the bottom of the openings, allowing for varying amounts of venting at the top and the bottom. Additionally, the high ceiling at the perimeter wall encourages hot air to rise and vent. The most innovative part of the system is that one of the operating panels in each bay is comprised of opaque plywood panels. Kahn alternated the default position of the opaque and clear panels so that when in their home positions, a dynamic pattern is created on the facade. Alternating positions of the panels also allows for various degrees of privacy and framed views. Kahn and Tyng developed interior elevation drawings for presentation to the Weisses that demonstrated alternate arrangements of the windows from inside the living room at day and at night (Figure 7.8).

The exposed rafters at the living room are built-up sections comprised of two 3 × 14 timber members spaced by a 2 × 10 that is flush to the underside of the exposed 1 × 6 tongue-and-groove roof decking. They are bolted together to form a composite beam with a profile that includes a reveal, bringing attention to the fact that there is more than one rafter at each bay. Use of a shadow joint in this case also compensated for irregularities of the timber dimensions that were part of assemblies fabricated in the field. Roof loads are transferred from the rafters to hidden corresponding posts that carry the roof load to the ground between

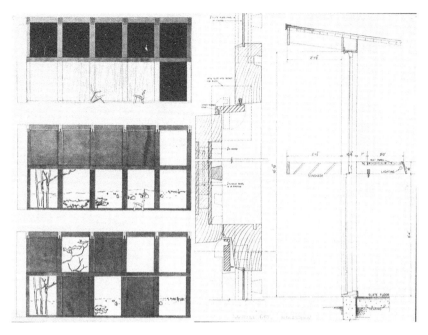

FIGURE 7.8 Drawings of different window panel options and schematic detail of the living room window wall system, Louis Kahn Architect.

Source: Louis I. Kahn Collection, University of Pennsylvania Architectural and Pennsylvania Historical and Museum Commission.

the windows. The hidden 2 × 6 wood posts correspond with the rafters above (Figure 7.9). Like the beams, they are separated by a 1–1/2-inch space. However, in this case the 1–1/2-inch gap allows for an outrigger to extend past at the roofline, and where there is a fixed horizontal shade, continue inside to support the light shelf.

The window jambs for the custom window frames are nested to the vertical posts and routed to receive a 1–1/16-inch parting strip that separates two channels for the window sashes (Figure 7.9). As a result, there is a significant gap between the two sashes, even when sashes are in their home position. To address this, Kahn implemented a rotating metal bar called a weather stop to close the gap when both sashes are returned to their closed starting position. When the top sash is in the lowered position, a lip on the top rail rests on top of the metal bar, which is positioned vertically, sealing the window. Operating window sashes are single glazed and opaque panels are comprised of two plywood sheets spaced approximately an inch apart and filled with ridged insulation. The fixed windows on the south side are double-glazed—something that would have been unusual at the time—especially in residential construction. Double glazing for the movable sashes would have resulted in an extremely heavy sash. At the exterior face of the

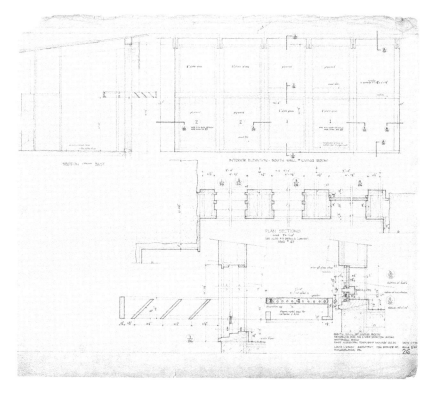

FIGURE 7.9   Detail sheet, Louis Kahn Architect.

Source: Louis I. Kahn Collection, University of Pennsylvania Architectural and Pennsylvania Historical and Museum Commission.

windows, Kahn layered two pieces of wood. The first is notched to receive the jamb, and a second layer of trim on the first is narrower so that an insect screen frame covers the first layer, resulting in a continuous outer face.

The depth of the outer face of the exterior louver is eight inches tall, corresponding with the depth of the outrigger that supports it. The shading device includes louvers that run perpendicular to the windows angled at 45 degrees and extend to the bottom and top edges of the outrigger. An effect is to render the louvers largely invisible from a distance, although some of the reflected light is visible on the facade. While the outer portion of the louvers was not novel, its integration with a novel interior shelf was. The shelf included radiant heating and electrical lights. Initially, the lights faced up and washed the high ceiling. In the final scheme, the lights were set under the shelf, providing more direct light to the area below. The shelf includes a 2-inch deep plaster layer at the top, poured within a steel angle lattice, with radiant heating pipes set into it. This complements a radiant heating system imbedded in the floor.

Tyng's influence on the project resonated beyond planning to detailing, with the equilateral triangular scupper shape at the ends of the roof valley (Figure 7.10).

FIGURE 7.10   Scupper and downspout at the east end of the Weiss Residence.

Source: Louis I. Kahn Collection, University of Pennsylvania Architectural and Pennsylvania Historical and Museum Commission.

The oversized triangular pans celebrate the capture of water at the roof valley ends, adding a level of poetry to an otherwise mundane building function.[20] They are the first example of a fundamental shape that Kahn used in subsequent projects as an accent, including the entry canopy of the Radbill Addition (1947–51) and the main stairs of the Yale Art Gallery. Tyng is also acknowledged to be the originator of the tetrahedron ceiling grid at the Yale Gallery, modeled after Buckminster Fuller's space frame experiments.

Conclusion

Morton and Lenore Weiss moved in at the end of 1949 and the residence was completed soon after in 1950, as Kahn was entering his fourth year of service at Yale. It was well received, winning a 1950 AIA Philadelphia Chapter Medal, and was featured in a 1950 issue of *Architectural Forum*.[21] As evidence of the positive reception of the house by its owners, Kahn was able to return several years after construction to complete a mural with Tyng in the living room. The owners spent the remainder of their lives in the house, which they loved.[22]

Kahn, along with his collaborators, was able to craft a residence that responded to the functional, environmental and emotional needs of its inhabitants. The form of Kahn's Weiss Residence was more nuanced than his later residential designs and arguably a better match for living. (Like his post-Rome institutional buildings, Kahn's post-Rome residential designs were comprised of primary shapes that made interior spaces subservient to exterior form objectives.) He was able to rectify asymmetrical volumes by balancing nuanced elements within an overall framework tied together with structural rhythm and a functional roof form. Furthermore, Kahn explored a language of architecture that avoided neutrality with respect to local traditions and material by contrasting the permanence of stone with the impermanence of wood. Kahn's use of stone during this pre-Rome period is consistent with ideas of monumentality and timelessness that he earlier explored in his writings. It is a more nuanced monumentality that supported performance in exchange for symbolism necessary for public structures.

Kahn's shift toward more fundamental forms after his Rome Prize Fellowship did not include shading device projections. Shading devices were associated with modernism, and Kahn was aware that the devices should not be uniformly applied on different exposures of a building, or when apertures are different heights. However, the search for timelessness and symbolism in architecture are not by themselves adequate reasons to omit shades.[23] The adoption of air conditioning in the United States and its dominance by engineers reduced the demand for shading devices across all building types as mechanical systems building absorbed significant portions of construction budgets.

Without the influence of Tyng, who transitioned out of his practice in the mid-fifties, Kahn's enclosure designs became less scientifically determined and more intuitive, and he sought more monumental methods of shading the sun. For situations where air conditioning wasn't feasible, or hot climates where glare was a concern, Kahn integrated solar shading into an outer skin, resembling a ruin, that enveloped inner enclosed volumes. This approach essentially absorbed the shading function into the building form, rendering it less obvious and specific.

Although it is tempting to compartmentalize Kahn's interests into specific periods, the introduction of small, horizontally sliding opaque panels into one of his last buildings, the Philips Exeter Academy Library (1965–71), is evidence of continuity of Khan's thinking about light, privacy and control. For the library, he designed custom study-carrels at its perimeter adjacent to small operable panels that, when moved, mitigate light and views. Kahn found a feasible location for the individual regulation of environmental conditions to be compatible with the monumental magnitude of a library. The fact that Kahn revisited his earlier window innovations when air conditioning had so thoroughly impacted building enclosures in the United States points to continuity in Kahn's thinking, as well as the merits of solar control in architecture. Now that monumentality is no longer in favor, and solar design is gaining interest, Kahn's modernist contributions are more relevant than ever.

Notes

1. Louis Kahn, "Monumentality," in *Architecture Culture 1943–1968*, ed. Joan Ockman (New York: Rizzoli/Columbia Books of Architecture, 1993), 47–54.
2. Sarah Goldhagen, *Louis Kahn's Situated Modernism* (New Haven, CT: Yale University Press, 2001), 49–52.
3. Carter Weisman, *Louis Kahn: Beyond Time and Style: A Life in Architecture* (New York: W.W. Norton & Company, 2007), 54–56, 58–59, 64.
4. Ibid., 66–68.
5. Thomas Leslie, *Louis Kahn: Building Art, Building Science* (Chicago, IL: University of Illinois Press, 2005), 8–13.
6. Kahn was not a particularly strong engineer–architect. Rather, he was more engaged in the expressive meaning of structure than its physics. See Goldhagen, *Louis Kahn's Situated Modernism*, 77–80.
7. Louis Kahn, "Talk at the Conclusion of the Otterlo Congress," in *Louis Kahn: Essential Texts*, ed. Robert Twombly (New York: W.W. Norton & Company, 2003), 47.
8. For a comprehensive history of mid-century solar house design, see Daniel Barber, *A House in the Sun* (Oxford: Oxford University Press, 2016).
9. Gropius situated a sun-room at the western side at his home in Lincoln, Massachusetts.
10. Robert McCarter, *Louis Kahn* (New York: Phaidon Press, 2006), 33.
11. It was also the first house he designed after accepting the teaching position at Yale.
12. George H. Marcus and William Whitaker, *The Houses of Lewis Kahn* (New Haven, CT: Yale University Press, 2013), 119–20.
13. David Brownlee, *Louis I. Kahn: In the Realm of Architecture* (New York: Rizzoli, 1991), 37.
14. The project north orientation of the building is 30 degrees west of true north.
15. Alexandra Tyng, *Beginnings: Louis I. Khan's Philosophy of Architecture* (New York: John Wiley & Sons, 1984), 137.
16. Klaus-Peter Gast, *Louis Kahn, The Idea of Order* (Basel: Birkhauser, 1998), 21.
17. Goldhagen, *Louis Kahn's Situated Modernism*, 58.
18. Marcus and Whitaker, *The Houses of Lewis Kahn*, 42, 120.
19. The introduction of regular and visible ordering units to Louis Kahn's architecture is likely due to Anne Tyng. See Marcus and Whitaker, *The Houses of Lewis Kahn*, 42.
20. Anne Griswald Tyng, ed., *Louis Kahn to Anne Tyng: The Rome Letters 1953–1954* (New York: Rizzoli, 1997), 35–37.
21. "Modern Space Framed With Traditional Artistry," *Architectural Forum* 93 (September 1950): 100–5.
22. Marcus and Whitaker, *The Houses of Lewis Kahn*, 7.
23. Kahn incorporated a movable panelized window system for the Pincus Annex of the Philadelphia Psychological Hospital (1947–51) that did not incorporate a shading system.

8

PROSAIC ASSEMBLIES

The Rich Pragmatism of Sigurd Lewerentz and Bernt Nyberg

Matthew Hall

The Swedish architect Sigurd Lewerentz left behind a body of uncompromising and mysterious work, spanning three quarters of a century until his death in 1975. His early works, which included collaboration with Gunnar Asplund for the Woodland Cemetery (1917–56) in Stockholm, established Lewerentz as a subversive classicist. Subsequently, his contributions to the 1930 Stockholm Exposition briefly aligned him with a functionalist approach, most evident in the stark cubic mass of the Social Security Administration Building (1928). Later, his design for the Villa Edstrand was painstakingly revised, stripping away traces of a heroic white machine for living and resulting in an idiosyncratic collage of steel and masonry (1933–37). Lewerentz's turn from external influence was fully sealed in his designs for the twin chapels at the Eastern Cemetery in Malmö of St. Gertrude and St. Knut (1943–44), which foreshadowed his later work.

After a brief retreat from architectural practice to concentrate on his company, *Idesta* (which produced aluminum and steel door frames, glazing systems and other components), his return to practice in the mid-1950s yielded in a small number of idiosyncratic projects establishing his reputation as an enigma. These late works influenced a younger colleague, Bernt Nyberg, to reinterpret and deploy his strict yet explorative philosophy towards materials and methods. Nyberg also serves as a key source for understanding Lewerentz's design intent. Their relationship from 1966 to Lewerentz's death in 1975 included a series of intense collaborations that changed the course of their architecture through the mutual influence between an old master near the end of his career—and the younger radical formulating and refining his values. Both architects acquired expertise through fresh insight on each new design problem, and while Lewerentz influenced many architects of the period, Nyberg can be seen as his most direct yet irreverent disciple. Nyberg's interviews with Lewerentz provide a portrait of an architect maturing through experimentation and invention in an effort to reject defaulting to normative ways.[1] The concerns of these architects went beyond building, but the decision making and expression was locked tightly within the discipline to the extreme of refusing to even speak of the intent behind the work, leaving only terse statements describing fairly obvious technical aspects. The buildings appear ripe with potential interpretations but lack overt reference.[2]

While the early phases of Lewerentz's career rest firmly within particular *zeitgeist*, he ultimately became distanced from most of his contemporaries, often operating outside of architectural conventions and current practice. In the wake of the rapidly changing, production-oriented building industry of Sweden's postwar welfare state, he consistently deployed architectural moves in direct tension with contemporary methods. Each detail was purpose built but unexpected given the evolution of his architecture. This suggests a progressive struggle for appropriate solutions, and unison between technical opportunity and an appropriate architectural language. Such a reading is direct, yet for decades scholars have drafted theories in search of deeper meaning, as if an architect simply solving problems is too lucid a narrative for such idiosyncratic work.

In the case of Lewerentz, it is probable that he was simply unintentionally poetic, an architect of open systems leaving the readings and interpretations to the observer. This notion, best summarized as prosaic poetics, is inherently contradictory and fits a reading of Sigurd Lewerentz's architecture. The prosaic in architecture may describe those criteria that all buildings must meet—the elements that they all have in common. The shedding of water, the admittance of light, the distribution of systems and structural solidity are the business of architecture. Through the application of standard typologies (door, window, column, beam, etc.), one can establish a basic lexicon that every building embodies. Lewerentz's entire body of architecture may be understood as an attempt to elevate our habitual understanding of architectural elements, undermining convention at every turn. By investigating Lewerentz's intent, refinement and influence, a clear philosophy can be brought into focus that invests fully in the balance between technical prowess and aesthetic desire, elevating normative elements to celebratory and often alien constructs.

The relationship between Lewerentz and Nyberg, and in particular a comparison of their late work, sheds light on a particular strain of Swedish architecture shifting from postwar building surge to production-oriented practice, seeking a specific and contextual language for architecture. As globalization blurred national and aesthetic identities and efficiency became the major driver for Swedish architecture, Lewerentz and Nyberg retreated into the discipline of architecture, where exclusive detailing and critical interrogation of normative spatial types became dogma. Outside of gravity and climactic forces, the social and political trends of the outside world were distanced and suspended. To accurately illustrate this, a brief summary of Lewerentz's half century of practice is warranted before the late work can be critically investigated and the genesis of Nyberg's mature philosophy can be adequately understood.

Lewerentz's Early Work and the Haven of Beauty

Lewerentz and Nyberg were among the many Swedish architects participating in the extensive civic projects of the postwar era, utilizing this venue to preserve the slow and attentive process of laboring over each detail and precisely dictating its execution. Claes Caldenby states that the gamut of Sweden's architectural production in the immediate postwar era lies between two poles: "the haven of beauty" and production-adapted design. The first is where Lewerentz and Nyberg operated, outside of convention, treating each design problem as a new context that subsequently demanded a unique solution. The latter represented the majority of architectural production at a time when architects were marginalized, as their professional role as advisors to the client and design solutions were adapted to the technical and economic demands of production. In a strongly utilitarian welfare society, the church stood out as the only building category where architecture was granted a certain "existential dimension." Caldenby continues describing Lewerentz as an "emblematic artist-outsider."[3] This description is apt, and while Lewerentz's late work is often described as a turn from his classical past,[4] upon closer inspection, his contrarian assemblies trace a clear lineage through his entire body of work.

Lewerentz practiced architecture from 1908 until his death in 1975. While most architects become less adventurous with age, his subtle subversion of existing typologies and common materials became more extreme. For Lewerentz, his radical nature would gradually reveal itself in each project as he aged through uncompromising formal and material unison. Concentrations of his built work were completed over decades at particular sites such as the Woodland Cemetery in Stockholm (1917–56) and the Eastern Cemetery in Malmö (1916–73). Both sites served as laboratories for continuous experimentation, where shifts in design approach through a reaction to a previous solution are traceable. Two structures in particular signal Lewerentz's abrupt departure from his classical forms.

At the Eastern Cemetery in Malmö, the twin chapels of St. Knut and St. Gertrude (1943–44), combined with a crematorium and the intimate Chapel of Hope (1955), form a disparate collage of aggregate program. Here his strategy enables details that carefully orchestrate the connections between differing assemblies, rather than rendering all surfaces similarly, as was common in his earlier architecture. At the twin chapels, an interior structural shell of golden Lomma brick is clad with thin slices of marble. The floors are covered with fragments of cork and wood and an entry vestibule of filigree wood strips that wrap from wall to ceiling. Lewerentz's later addition of the Chapel of Hope transformed a coffin-loading dock into a third, smaller yet more intimate chapel. Here all surfaces are clad in linear marble slabs, scraps and wooden strips. The wall's true make-up is only apparent at openings, where the layers are revealed with concrete filling in the resultant gaps of contrasting masonry skins (Figure 8.1).

At his other life-long laboratory, the Woodland Cemetery in Stockholm, Lewerentz was tasked with creating a restroom addition (1956) to his 1925 masterpiece of "Swedish Grace," the Chapel of the Resurrection. Rather than connecting the new structure seamlessly with the existing complex of severe and abrupt classical volumes with stucco surfaces on the exterior, he designed a structure that was clearly distinct. Lewerentz stacked bricks with bleeding joints showing the back, or "bad side," turned outwards, showcasing manufacturing marks and unfinished surfaces, akin to a masonry wall awaiting rendering, as if the Chapel of the Resurrection was stripped down to its structure (Figure 8.1). These humble volumes are another example of subjugating gestural form to a high resolution of aggregate pieces and thus a final divorce from rendered surfaces. These two small projects, completed at roughly the same time, expose early experiments that would later

FIGURE 8.1   Chapel of Hope ceiling and window at the Eastern Cemetery, Malmö, and restroom addition to the Chapel of the Resurrection with masonry detail at the Woodland Cemetery, Stockholm.

Source: Photographs by Matthew Hall.

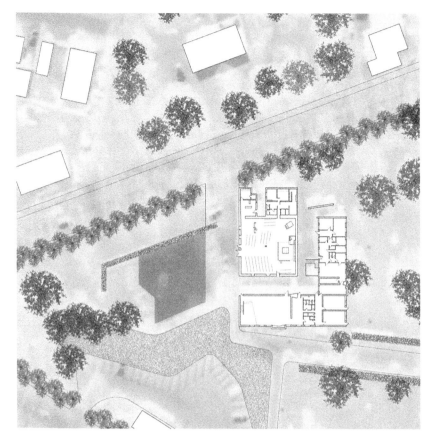

FIGURE 8.2   Plan of the Church of St. Peter in Klippan, Sigurd Lewerentz, 1966.

Source: Drawing by Matthew Hall and Timothy Smela.

formulate his attitude towards veneer brick, setting the stage for the first of two masonry churches, St. Mark's in Björkhagen, completed in 1960, and later the Church of St. Peter in Klippan in 1966 (Figure 8.2).

Enter Nyberg

In 1955, Bernt Nyberg completed his education at the Royal Institute of Technology in Stockholm. His practice began when he and Karl Koistinen, while employed at Klas Anshelm's office, won a competition entry in 1957 for a sports hall in Fagerstra. While the building was never built, the events led Nyberg and his first architectural partner to begin a seven-year collaboration that would serve as the most prolific period of Nyberg's career. Nyberg and Koistinen treated each commission with meticulous care and attention, developing a sense for detailing for masonry construction that would continue to be refined in the later work of both architects. Koistinen spoke of influential trips to study the work of Aldo van Eyck and Arne Jacobsen, which likely influenced their extensive work in Eslöv and Hörby, which consisted of education and housing facilities for the mentally ill.[5] After the pair split to form their own offices in the late sixties, Koistinen proceeded to complete a diverse and prolific body of work. In contrast, Nyberg achieved considerably less output and concentrated on fewer projects. It was not uncommon to have one design on the boards at a time, consuming the entire office and dedicating years to the production of a single building.

In 1964, a young and enthusiastic Bernt Nyberg met Sigurd Lewerentz on the construction site of St. Peter's Church in Klippan, Sweden, Lewerentz's last major commission. The pair became close, and late in the project Nyberg would often attend to the duties on site when Lewerentz was ill. Shortly after the completion of the project in 1966, Nyberg begin two projects that are clear responses to his experiences working with and documenting the process of construction in Klippan. The first was an addition to the Lund regional archives in 1970, and subsequently a funeral chapel in Höör in 1972 (Figure 8.3) that would be his first and last religious project. While he was reconsidering his ideas about architecture in discussion with Lewerentz, Nyberg was able to develop his own unique, but related, language. The architects posited similar questions resulting in radical responses, and within six years both architects would be dead.[6]

While Lewerentz and Nyberg completed other works after St. Peter's Church in Klippan (Figure 8.4) and the Funeral Chapel in Höör (Figure 8.5), these two spiritual spaces provide the best representation of Lewerentz and Nyberg's converging values while simultaneously exhibiting where they differ. Given the richness of these projects as assemblies of thought, they may be best understood through dissection. A taxonomy for parsing these works begins with the smallest unit—the brick. It extends to common relationships that works of architecture share: column to beam, window to wall and the consequences of negotiating the needs of interior and exterior requirements.

Brick: Combination and Composition

When combined, the units of a masonry assembly enable new readings that transcend the possibilities of individual bricks. However, there is a difference between assemblies that are the result of combination, as opposed to composition. Combination is generally the ad-hoc assembly of materials via a value system rooted in pragmatic aims. Composition suggests extending beyond mere problem-solving and quotidian solutions.

For postwar Sweden, brick was the predominant material, and near the end of the massive building boom of the fifties, there were close to 30 brick factories in the southern region of Skåne alone. Brick was economical, and its manufacture and installation in load-bearing walls put thousands of masons to work.

FIGURE 8.3   Plan and section of the Church of Funeral Chapel in Höör, Bernt
Nyberg, 1972.

Source: Drawing by Matthew Hall.

By the early seventies, brick production was reduced by half, and within ten
years it plummeted, leaving only five factories in the region.[7] In addition, government funding for public architectural projects was becoming more limited.
With these developing circumstances, Swedish architects were shifting away
from load-bearing construction to more efficient veneers and layered assemblies.
Lewerentz and Nyberg would both respond to the current trend of reduced brick
production, and its increased cost, with hybrid assemblies of reinforced concrete

FIGURE 8.4 Interior of the sanctuary space, Church of St. Peter in Klippan.

Source: Photograph by Matthew Hall.

and masonry. They embraced the challenge to find new techniques for achieving massive walls by injecting the assembly with more than just structure and enclosure. In partnership with mechanical engineers at BACHO, both architects integrated purpose-built mechanical systems that integrated with the mass of the

FIGURE 8.5 Interior of the Funeral Chapel in Höör.

Source: Photograph by Matthew Hall.

building, allowing air to disperse through cavities in the walls. This thickness of the wall system also facilitated acoustic control by removing mortar in vertical slots to absorb sound via hidden mineral-wool panels.

Both Lewerentz's Church of St. Peter and Nyberg's Funeral Chapel in Höör deployed bricks produced at the *Helsingborg Ångtegelbruk*, which opened in 1873 and ceased production in 1978. The brick used measured 250 × 120 × 65 mm. Lewerentz had a custom brick produced with a specific machine that pressed a surface quality and darker color, perhaps due to its hard, burnt characteristics. Nyberg used the standard Helsingborg brick of the same dimension, yet the architects had radically different approaches to the methods of construction. The surfaces at Lewerentz's church are fluid as the mortar joints expand and contract in running bond. The strict rule to never cut a brick forces the mortar joints to mediate the strange geometries and interruptions in the surface. Lewerentz was concerned with each brick expressing precise patterns, textures and orientations, in some instances with a focus on a single masonry unit. The construction specifications for St. Peter's Church state that the surface layer of bricks "are distributed according to the architect's closer notifications during the work process." This supports numerous accounts of Lewerentz constantly being on site directing the masons first hand, often down to the placement of specific bricks. The specifications continue on to describe the intent for masonry work in dry technical prose, with one exception: "there will be some local deviation from the strict masonry, deviance, gives life to the surface."[8] Along with this, there is a note stipulating leaving each brick in the place it was initially laid with no disturbance, with an allowance for up to a centimeter's digression. An encouragement to vary the vertical mortar joint width further explains how the quality of the wall was achieved. Lewerentz was enamored with the archaic ruins of past civilizations, where a lack of precision in hand-formed bricks necessitated wide mortar joints to adapt to the inconsistencies in the masonry units. This inclination is evident in the bond of St. Peter's, where the overall wall and opening dimensions do not correspond with standards expected for masonry, resulting in wildly varying joints (Figure 8.6). Lewerentz desired a surface with "life" but sought ways to predicate

FIGURE 8.6 Masonry comparison, St. Peter's Church (*left*) and Funeral Chapel in Höör (*right*).

Source: Photograph by Matthew Hall.

its realization through a rigorous set of demands, no doubt challenging for even the most competent mason.

In the worship space of the Church of St. Peter, Lewerentz allowed the brick to cover all surfaces, defying masonry's normative structural conditions as if the brick were featherweight rather than stone-like. For example, the floor surface of the worship space undulates and fractures at the baptismal font, interrupted by lines of concrete that dictate chair arrangements. It is notable that both Lewerentz's and Nyberg's projects utilize individual chairs rather than the heavy linear pew, further emphasizing that the interiors and exterior details rely on the aggregation and variation of repetitive elements. It is notable that in the adjacent administrative and commons building of St. Peter's that masonry is relegated to only walls, leaving tile and wood for floor and ceiling—a classic dichotomy between sacred and profane.

At the Funeral Chapel in Höör, Nyberg would take the opposite approach. For him, the wall was about totality, and he would often ask, "Why does the wall need to be about the single brick?"[9] Where Lewerentz's bricks float and swell as if they are a thin viscous fluid on walls, floors and ceilings, Nyberg's masonry obscures the boundaries between bricks, allowing the projecting mortar to read as the primary building surface material (Figure 8.6). The process involved buttering the brick, and once laid, the compression naturally results in overflow. Rather than tooling and cleaning the joint, the masons were instructed to scrape the surface with a wooden block, resulting in a blurred surface. The wall transforms into a landscape when struck by sharp light, casting shadows and providing a sense of depth. For Lewerentz and Nyberg, the mortar balances its own expressive identity in codependence with the individual units of masonry. Both architects had stringent yet contrasting ideas regarding masonry construction, taking ownership of the unit and contextualizing it to a unique application.

Contradiction and Opportunity

For Lewerentz and Nyberg, the extent of their use of masonry became limited compared to many of their contemporaries. For Lewerentz, the brick became thin: a single wythe at floor and ceiling layered over a combination of concrete and utilitarian brick structural bearing walls. For Nyberg, the relegating of structural brick to vertical elements allowed for tension and contrast with the varied materials of the horizontal surfaces of floor and ceiling. Both buildings deploy wall assemblies that defy normative readings of load paths. What appears to be massive is often hollow, enabling air distribution from mechanical spaces below to cracks at windowsills or depressions and voids in the walls. In addition, the wythes experienced are usually a veil, masking what lies beneath.

Nyberg and Lewerentz were contradictory at every turn. Sometimes they were hell bent on expressing the absolute clarity of exposing weld joints or material thickness. In other cases, they were interested in erasing any notions of load transferring lintels or readings of mass. For Lewerentz, these tendencies are derivative of his earlier brick church in Björkhagen, the Church of St. Mark (1960). Early insight into Nyberg's penchant for illusion and contradiction can be found in his approach to an addition to the 1905 Regional Archive in Lund. The original wing designed by Carl Möller was broken down with the usual moves that define scale and proportion to give evidence of occupation, while Nyberg creates a seemingly scaleless mass with no hint of the structure's purpose or interior. Original building and addition kiss at glass panels, broken by the stair and elevator core, concentrating attention to one delicate joint. The addition has all the signs of load-bearing masonry construction, as bricks return the edges of deeply set openings appearing monolithic, yet its surface is comprised of a single wythe thick, with no expansion or control joints furthering the illusion of mass. The building's enclosure is a stage set, but only an observant visitor schooled in the practice of masonry would see through the deception.

Unique to both projects is a hybrid structure of masonry, concrete and steel. Both architects orchestrate a highly internalized approach that presents us with brick mass and then reveals steel structures locked within the major spaces of the projects. These structures relate little to any standardized notions of steel construction. They are not archetypal, but distort the roles of column and beam while reducing the structure to an object rather than a typical frame and grid. Expectations regarding the meeting of wall with ceiling and column to beam produce a tension between artful structural expression and compositionally ad-hoc solutions. The structures appear like a prop set in place to prevent collapse. In Lewerentz's sanctuary space, the drama of alternating vaults is complemented by a pair of weathering steel wide flange columns. The structure was supposedly produced and left outside to weather for six months, resulting in a consistent and deep rusty patina at odds with its interior condition. The twin column meets horizontal partners that sprout miniature cruciforms like tiny branches to meet the steel sections at each vault's peak and valley (Figure 8.7). Why would Lewerentz design a structure that places a massive column dead center in the space? It appears as if the span was simply too long for the brick vaults and the prop was needed: but could more columns be used to create aisles with a centralized nave? In the wake of any clear reason, the structural act is on forceful display.

Nyberg had a rule to never use a hollow section, always presenting the true edge thickness of steel members, motivated more by a desire for a more complex visual effect than any notions of being honest. Nyberg was quoted as often seeking to "get something for free,"[10] like the edge of a waffle slab cut mid coffer producing a highly articulated edge profile. This doctrine, combined with a tendency to facilitate reveals and joints, produces a column that is simultaneously bulky and thin, massive but filigree. The blossoming cruciform columns in Nyberg's chapel support a coffered concrete slab held in precarious stasis on thin stilts of steel. These columns are aggregated from small angles separated with steel spacers, exacerbating the vertical nature of the columns and their relative

FIGURE 8.7 Steel structure comparison, St. Peter's Church (*left*) and Funeral Chapel in Höör (*right*).

Source: Photograph by Matthew Hall.

thinness compared to the massive concrete ceiling they support (Figure 8.7). In turn, massive walls laterally brace a perilously floating waffle of concrete on thin steel spines that appear to stretch in tension. At the subtle crack between concrete ceiling and massive wall, the brick abruptly shifts to aluminum foil with no change of plane. The foil reflects light from above, picking up a tint of blue from paint on the waffle slab's edge. This effect is then amplified by the warm color temperature of the incandescent bulbs occupying the coffers of the waffle slab. No matter how grey the days are in Skåne, one is guaranteed blue skies at their funeral. Like many of Nyberg's illusive details, the design orchestrates a particular effect, but it is never a true deception, as clear hints are provided to decode as the sequence of the space unfolds. Unlike Nyberg, Lewerentz's illusions remain intact and evolved into speculative myth. This pair of architects could very well have solved their structural issues though massive construction, but that would be a missed opportunity, antithetical to their process and philosophy.

The Openings

With Lewerentz's late architecture, openings are devoid of frames, juxtaposing fragile glass directly with massive walls. This strange technique is further compounded by the fact that Lewerentz held patents for numerous door and window frames, and had been producing them under the name *Idesta* for decades. In his final two churches, St. Mark's in Björkhagen (1960) and St. Peter's in Klippan (1966), he jettisons the window frame, experimenting with glass panel placement on the interior at St. Marks, and then transitioning to the more direct detail of hanging the insulated glazing units on the exterior surface at St. Peter's. For the

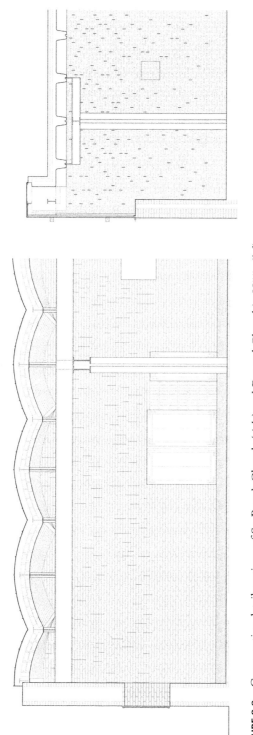

FIGURE 8.8 Comparative detail sections of St. Peter's Church (*right*) and Funeral Chapel in Höör (*left*).

Source: Drawings by Channing Broadie.

former, the building appears to be ruin from the exterior, glazing set deep within the mass of the wall. This detail required a complex copper sill to shed water and protect the wall assembly. For the latter in Klippan, the insulated glass panel is simply clipped to the exterior and sealed with Tremco Mono, an early black silicone glazing compound. Here, the thickness of the insulated glass is on clear display, and from the interior there is no evidence of enclosure provided as the wall appears to lack any glass at all (Figure 8.9). When asked by Nyberg, Lewerentz simply replied that it was much simpler to do it this way, continuing to state that it was not something that should normally be done and suggesting that this was a special case.[11] There was no deep philosophy, no treatise on the expression of cutting-edge building technologies of insulated glass and sealant; it was simply a solution to the problem of a window in a wall, a simple problem that all architects must resolve, and one with which Lewerentz spent over 50 years experimenting.

Nyberg's thinking on the openings at the chapel in Höör is analogous to Lewerentz, but is solved with an entirely different approach. Here, Nyberg recesses the exterior glazing flush with the exterior masonry surface with projecting weathering steel spines at the butt joints. Attached to these spines are angle-shaped clips to secure the masonry as the structural glazing sealant cures (Figure 8.9). Nyberg chooses to make these clips permanent, amplifying their presence by extending the projections well past what is needed. The effect of these miniature projections has an enormous impact on the otherwise flat surface. They are only articulated by the subtlety of the projecting smeared mortar joints and razor-like projections that catch light. This is yet another example of Nyberg's tactic of "getting something for free," where a necessity becomes an opportunity for articulation and expression.

Like Lewerentz's strategy at St. Peters, the walls of Nyberg's chapel are hollowed out for the passage of air. In addition to creating venting penetrations in

FIGURE 8.9 Glazing detail comparison, St. Peter's Church (*left*) and Funeral Chapel in Höör (*right*).

Source: Photograph by Matthew Hall.

the interior masonry surface, he deploys air passageways at windowsills, similar to Lewerentz's solution in the administrative wing of St. Peter's. Given the difficulties in predictive thermal behavior of the window assemblies at the time, Nyberg planned for a secondary internal piece of glass flush with the interior surface, trapping warm air in the cavity between interior- and exterior-facing glass. The insulated units, however, proved to be effective, but the framing for the secondary glass was left in place as a record of process or a ruin to be harnessed with some future glazing technology. From discussions with Nyberg's former collaborators, this was an intentional move. It was already in place and added another layer of potential readings to the wall, so why not leave it?

Combination Before Composition

Per Olof Olson sums Lewerentz up well in the following statement:

> He seems to be more interested in technical and constructive problems than esthetical, and is often shocking with "ugly" architectural solutions. He belongs to the few innovators to whom ugly or beautiful doesn't exist. It's all about finding an obvious solution to a problem. . . . Sigurd Lewerentz is a realistic technician and open-minded esthete.[12]

This description is also apt for Nyberg. The architects' efforts epitomize an ethic that focuses on the problems of building, but not in the purely utilitarian sense of the word. Both architects were clearly interested in form. For Lewerentz, the form is legible through an aggregation of diverse solutions that often blur and overlap; for Nyberg, the form is divided into its requisite parts, all identifiable as normative elements but translated and preserved, maintaining their autonomy. In the case of Lewerentz's late brick churches and Nyberg's funeral chapel, the formal approach can be read as a sum of multiple and often contradictory assemblies held in a tense stasis.

While some architects play with the composition of form, Lewerentz and Nyberg played with the combination of materials. An understanding of their work offers us a unique example of two generations of postwar European architects struggling to articulate an appropriate language and architectural identity for the rapidly changing cultural context of a developing European nation. Surviving collaborators of both architects often recall the immense joy that they found in the consistent and critical interrogation of convention. For these architects, there are no clear answers, but the richness of their thinking continues to infatuate, open to interpretation and ripe for potential readings.

Notes

1. Lewerentz, Sigurd, in discussion with Bernt Nyberg and Karl-Erik Olsson (photographer), February 1967. In a discussion with Bernt Nyberg, Lewerentz remarked disparagingly against a functionary who only conceived of one correct and absolute way of doing things, stating with rare eloquence, "He had never understood

Matthew Hall

that you can do it in a hundred different ways. That is what we must acquire. Knowledge." In both Nyberg and Lewerentz's case, knowledge required practice, experimentation and dialog, and they chose to exercise all three silently through the act of building.

2. Bernt Nyberg's narrative for the Chapel in Höör is brief and to the point: "The Chapel at Höör is the product of an open architectural competition. The original program was in connection with a burial chapel. However, the discreet manner in which the handling of coffins was solved provided the congregation with the opportunity of inaugurating the building as a chapel which was open for all forms of church activities. Among other things the architectural idea involves two square roofs, one inside and one outside of the building. The plain lantern-light glass in the light openings of the roof is simple windscreen glass with electrical wiring for the thermostat controlled snow melting and condensation control. The other glazing is the form of thermo-panels."

3. Claes Caldenby, "Lewerentz and the Haven of Beauty," in *Lewerentz's S:t Petri at 50: Context, Fragments and Influence*, eds. Matthew Hall and Hansjörg Göritz (Klippan, Sweden: Municipality of Klippan, 2016), 47–53.

4. Wilson Colin St John, "The Sacred Buildings and the Sacred Sites," in *Sigurd Lewerentz 1885–1975*, eds. Nicola Flora, Paolo Giardiello, and Gennaro Postiglione (Milan, Italy: Electa Architecture, 2001), 11–34. Colon St. John Wilson describes this turn as "extreme, unblinking, and absolute. His classicism was more refined, more deeply felt, more original than that of any of his contemporaries; his late work was more austere than any minimalist, more uncompromising than any Brutalist."

5. Karl Koistinen, "Karl Koistinen Bernt Nyberg Arkitektkontor," in *Endangered Architecture, the work of Bernt Nyberg*, ed. Matthew Hall (Copenhagen: Skissernas Museum in Lund Sweden, Lund University and The Royal Danish Academy of Fine Arts, School of Architecture and Technology, 2015).

6. Lewerentz died of natural causes in 1975 in Lund, Sweden. Around the same time, during a trip to Japan, Nyberg was diagnosed with multiple sclerosis. Treatment was unsuccessful, and after three pain-filled years, he committed suicide in 1978.

7. Tomas Tägil, Tomas Gustavsson, Kristina Bergkvist and Björrn Magnusson Staff, *Modernismens teglfasader* (Arkus Skrift nr 65 Stockholm, 2011), 81.

8. From the construction specifications for St. Peter's Church in Klippan: Structural back walls shall be 1,6 perforated brick according to the regulations and construction documents and shall be approved by the structural engineer and architect. At cladding masonry with running bond, the courses are encouraged to vary in width—equilibrium can be done every 3–4 shift. Single stones can gladly digress a centimeter from the flush wall. Deviance, gives life to the surface. The significant is that the brick walls are largely left in vertical with the sinker, and the masonry niches and endings, are at the right height. An important detail in the work operation is that when a brick has been inlaid to the mortar, it shall not be displaced from its location by taps, etc.

9. Staffan Schultze (architect and close collaborator in the office of Bernt Nyberg), interview by Matthew Hall, December 2014.

10. Per Iwansson (architect and close collaborator in the office of Bernt Nyberg), interview by Matthew Hall, December 2014.

11. Per Iwansson, narrative and translation of interviews with Sigurd Lewerentz conducted between 1965–70 by Bernt Nyberg, *Lewerentz's S:t Petri at 50: Context, Fragments and Influence*, eds. Matthew Hall and Hansjörg Göritz (Klippan, Sweden: Municipality of Klippan, 2016), 111–15.

12. Per Olof Olson, "The Lewerentz Phenomenon," originally intended to be published in the first monograph on Sigurd Lewerentz, ed. Bernt Nyberg, 1974.

Translation assistance: Andreas Förnemark.

9

"THE MATERIAL OF THE FUTURE"

Precast Concrete at the 1962 Seattle World's Fair

Tyler S. Sprague

From the very beginning, the US Science Pavilion at the 1962 Seattle World's Fair (Figure 9.1) was more than a building. Those associated with its design and construction—including the architects, engineers, builders, fair organizers and federal government officials—all described an ambition and significance for the Pavilion that far exceeded its modest footprint. The US Science Pavilion was conceived as a building of the "future" and charged with embodying the characteristics of an idealistic, science-oriented, near-present world. Those involved were collectively challenged to contemplate and realize the future of architecture, structure, material and public space.

Within the deliberation surrounding the design of the Pavilion, one material was nearly unanimously selected as the primary building material: precast concrete. At a time when advanced plastics, high grade steel and aluminum were being experimented with at architectural scales, the familiar, rather unassuming concrete was vaulted to the forefront of futuristic design. A basic mix of cement, aggregate, sand and water: concrete had long been used in architectural and infrastructural projects. Now, the precasting, and prestressing of embedded cables, was fundamentally changing the textures, forms and structural capacities possible in concrete. Acknowledging the impact of these new techniques, the design of the US Science Pavilion would employ design qualities in direct contrast to earlier associations of concrete. The Pavilion aspired to be light, thin and gleaming white, a far cry from the heavy, bulky and uncompromisingly grey concrete of earlier times. The future would bring new understandings, not through substantially new materials, but through new conception, execution and technique of a familiar material.

Despite lofty aspirations, the design had to be executed by a team of individual specialists, within very practical programmatic, financial, technological and

FIGURE 9.1 Aerial view of US Science Pavilion.

Source: Courtesy of Washington State Digital Archives.

time-based constraints. The constellation of people assembled for this project were centered in Seattle, and unified by their interest in expertise in precast concrete. Together they defined the US Science Pavilion as a project unique to its time and place. Each collaborator brought something different to the process, including vision, experience, willingness to take chances, compromise, meticulous attention to detail and perseverance. Through these eternal human traits, local actors become facilitators of precast concrete as a technology, material, architecture and agents of the future.

With few limits other than imagination, the future held the potential for brilliant new directions of human life. But it also warranted reconciling with permanence, and a more somber acknowledgment of time. While the future was bringing radical change, the present buildings that lasted would become part the future themselves. Through its use at the US Science Pavilion, precast concrete came to fit this nuanced vision of the future (Figure 9.2).

1962 Seattle World's Fair

The 1962 Seattle World's Fair proclaimed itself to be "America's Space Age World's Fair," and Seattle, a rapidly expanding city, was an appropriate setting.[1]

FIGURE 9.2   US Science Pavilion at Night.

Source: Courtesy of Seattle Municipal Archives Digital Collection.

The Boeing Company's jet engine-powered 707, released in 1958, was transforming air travel nationwide, making destinations (like Seattle) more accessible to the eastern parts of the nation. Meanwhile, national ventures into outer space were taking on more international significance. In response to the launch of the Russian Satellite Sputnik in 1957, President John F. Kennedy delivered a speech to a joint session of Congress on May 25, 1961, calling for an ambitious space exploration program.

Named the Century 21 Exposition, buildings and fairgrounds were intended to point towards the future through an exciting display of technological innovation. After the Fair was over, the city anticipated turning the grounds to a permanent park or civic center. The federally funded US Science Pavilion was to be one of the largest complexes on the Fairgrounds. The Fair promotional material stated that the US Science Pavilion "will be a large, ultra-modern building designed to reflect the concept of its contents—the world's scientific vistas" and to express "the role of science in the Space Age."[2] The Pavilion would contain several designed exhibits, showcasing the role of science in visions of the future.

The federal government also wanted the possibility that the Pavilion buildings could be used for another purpose after the fair was over. Lacking a specific vision themselves, the GSA (General Services Administration) asked prospective architects to consider secondary use as an office building. During the fair, it was to be a compelling exhibition. After the fair, it was to be a common government office building. In short, they were looking for a building both dedicated to the exciting future that lay ahead and able to support the mundane, generic activities of present-day office work. Attempting to clarify this discrepancy in a letter to prospective architects, the GSA described a responsibility to "build an Exposition Building with consideration for secondary use as an office building, not an office building for temporary use as an Exposition Building."[3] With this priority, the architects should speculate on potential futures, and the actual program would materialize.

Minoru Yamasaki

Yet the future use of the Pavilion was an important architectural issue. In 1959, after a survey of local and national firms, the federal government awarded the design of the US Science Pavilion to Minoru Yamasaki (1912–1986).[4] Yamasaki was raised in Seattle and earned his degree in architecture from the University of Washington in 1934, before practicing in New York City and establishing his own practice in Detroit, Michigan, in 1949.[5] With a rising national status, Yamasaki was selected based on his experience with large exposition buildings and enthusiasm for returning to Seattle.[6] To execute a project there, Yamasaki partnered with the local architecture firm Naramore, Bain, Brady & Johanson (Figure 9.3). Yamasaki was classmate of Perry Johanson at the University of Washington and the two remained close.

Yamasaki soon understood the dual nature of the US Science Pavilion project as both an exciting Fair building and a permanent part of Seattle's civic center. Shortly after his appointment, Yamasaki took issue with the federal government's plan for future conversion. In a letter to the GSA, Yamasaki wrote:

> The further we look into this question the more impractical it appears to us. This building, whose basic purpose is to tell the vastly interesting story of contemporary science, if compromised for a future obviously different use will not be successful in its original use.[7]

The incompatibility of the two conceptions of the future threatened the entire project. His office assessed the feasibility of transforming the Pavilion to support four different uses: a warehouse or garage, an office building, a scientific research center, and a permanent science or other museum. For Yamasaki, of these four options, only the last one, which was similar to the intended fair use—was found to be even remotely viable. For both practical and conceptual

Property of Museum of History & Industry, Seattle

FIGURE 9.3   Looking at the US Science Pavilion model, from *left* to *right*: Mr. William Ku, Assistant to Mr. Yamasaki; Mr. Perry B. Johanson, Architect of the firm of Naramore, Bain, Brady & Johanson; Mr. Philip M. Evans, Commission, US Science Exhibit; Mr. Francis D. Miller, Deputy Commissioner, US Science Exhibit; and Mr. Minoru Yamasaki, Architect.

Source: Courtesy of Museum of History and Industry.

reasons, he believed the future use of the building should maintain the original charge, of celebrating the cultural legacy of science. Despite the urge to design for a completely adaptable building, Yamasaki articulated a confidence in a continuity of use.

> Since the United States sponsors a marvelous institution of the East Coast, the Smithsonian, to record our countries fabulous history of technological and scientific progress would it not be logical to sponsor a similar though smaller institution on the West Coast?[8]

The government soon agreed and began to make arrangement for a future science center. With these priorities set, Yamasaki shifted his thinking to the material and spatial arrangement of the Pavilion. Considering occupancy, the need for a fire-proof structure and permanence of the building, Yamasaki recommended concrete with an "aggregate" surface as an economical option.[9] His simple suggestion of concrete on the grounds of cost satisfied the practical considerations of the GSA, but also covered his much more in-depth understanding of the material. Concrete—specifically precast concrete—would soon carry nearly all of Yamasaki's architectural ambitions.

The US Science Pavilion Design

Yamasaki had experimented with expressive concrete on previous projects, including the thin-shelled St. Louis-Lambert Airport (1954) and the folded-plate ACI Headquarters (1958). But on this occasion, he was more interested in using his architectural design to create a peaceful setting. Yamasaki wanted to harness the advancing technology and use it to create a human-centered experience. In his evolving perspective on modern architecture, Yamasaki stated that he aimed to create buildings to "love and touch."[10] He wanted his Pavilion buildings to have a sensitive, tactile quality. Yamasaki was interested in combining the high-quality texture and finish of precast concrete with the thinness of shell construction, to create a serene environment.

With several large buildings already planned for the Fair (the Washington State Pavilion, the Civic Auditorium and others), Yamasaki was not interested in designing yet another "large structure." Yamasaki instead chose to design several small buildings instead. In addition, based on his experience visiting previous World's Fairs, Yamasaki felt the implied "competition" between different buildings created "an impression of architectural chaos." To contrast this, Yamasaki chose to design with equal attention to inside and outside space.

Yamasaki was also interested in designing a vertical tower to punctuate the visual presence of the pavilion. Yet similar to his strategy with the buildings, when Yamasaki heard of plans for a 600-foot tall Space Needle on the fairgrounds, he decided to design five smaller towers arranged around exterior fountains and walkways instead. As a result of these objectives, Yamasaki designed a central courtyard space as the primary focal point of the Pavilion. This space was to be surrounded and somewhat enclosed by several, separate buildings. Each building would be a simple, rectangular assembly of exterior, load-bearing precast panels, and a long-span beam between. End panels would enclose the box-like structure. The buildings would have no interior columns, leaving wide-open floor plans for exhibits of different sizes. The walls would be articulated by a pattern of tightly spaced columns or studs, only three feet apart, giving the walls a "pin-striped" appearance. Curving ribs trace out pointed arches in between the columns. The flat wall surface in between the columns would be solid in some places and open in others—allowing for colonnade-like walkways and windows in different locations.

The towers would consist of four thin columns, each supporting an open, domed lattice work. Each tower was composed of four supporting corner columns topped with connecting arches that rise higher than the surrounding buildings.[11] The arches came together to create an open, dome-like shape, filled in with a thin tracery similar to the pattern of the building walls below. Despite their delicate appearance, the towers were to be made entirely of precast, prestressed concrete. Each individual piece was to be brought to the site and assembled together in place.

With thinness held paramount, Yamasaki designed the pavilion walls as both enclosure and structure. In his styling, Yamasaki gravitated towards slender ribs and pointed arches, often described as a "Space Age Gothic." Yamasaki stated:

> The Gothic motif was not deliberately Gothic, but after having tried several designs in which we tried to achieve an uplifting and soaring feeling, we felt that this design for the towers gave us the utmost feeling in this direction.[12]

Thinness and lightness were held paramount in the design, in order to create an uplifting and soaring feeling, Yamasaki wanted to keep all of the building elements as small as possible. The visitor would enter the space via a series of elevated platforms, with soaring arched towers overhead, fountains and pools below, contained by the gleaming facades of the surrounding buildings.

For Yamasaki, the use of precast concrete was essential to achieve this architectural goal—on both practical and conceptual ground. Working within a tight timeframe, the off-site production and rapid construction of precast panels would greatly accelerate the Pavilion's schedule. In addition, buy using the same material for all buildings (and towers), precast would become a unifying element among the separate parts offering economies of scale. Yamasaki knew that precast concrete could achieve these architectural goals, but the execution was far from certain. Precast concrete, in the early 1960s, was still in its infancy in the United States—requiring local expertise. Yamasaki would need to rely on the technical capabilities of the Seattle-area building community.

Concrete in the Pacific Northwest

Fortunately, innovation in precast concrete was uniquely suited to the Pacific Northwest, which already had a legacy of creative design in concrete. In the 1920s, as a response to the high price of structural steel, local builders and engineers used concrete as a local, economical option. Large mineral deposits in the North Cascades were well suited for the production of local, high-quality cement, and dense, glacially deposited aggregates further enabled an inexpensive yet high-strength concrete. This inexpensive material led to a boom in reinforced concrete structures in the late 1920s.

In the process, the Pacific Northwest developed a nationally recognized expertise in concrete. Local builders and engineers pushed new code allowances, and local architects began designing with concrete in inventive ways. In the 1930s and 1940s, local figures like architects A. W. Gould and John Graham and engineers Homer Hadley and George Runciman produced several signature works. The Exchange Building, a downtown Seattle skyscraper, was one of the largest concrete buildings in the country when built. Hadley designed an innovative, floating, concrete pontoon bridge across Lake Washington.

However, this first wave of concrete innovation was based on a limited set of construction techniques that produced a particular type of concrete structure. The concrete mix of cement, aggregate and sand was mixed with sufficient water to become fluid enough to be poured into formwork. Steel reinforcing bars were placed in the formwork, prior to casting, becoming completely embedded in the final shape. These steel bars (solid, static) were designed take all tensile forces that developed within the overall structure. This simple arrangement allowed for the entire process to be completed on-site, with the construction of formwork, placing of reinforcing steel and pouring of concrete taking place at the job site. Concrete's poor resistance to cracking often resulted in large, sometimes oversized sections.

But as the 1940s turned into the 1950s, prestressing emerged as a new technique in the creation of concrete elements. The prestressing of concrete, while relying on similar materials as traditional concrete, required a completely new process of production: the pre-tensioning, or post-tensioning, of steel cables that are then cast within concrete. This process requires the stretching of high-strength cables, resisting thousands of pounds of force in a controlled setting. The tensioned cables, once released, force the concrete into compression, counteracting the splitting tensile stresses common in concrete. With tight control over the placement of cables and large jacks needed, the precasting of concrete in an off-site facility became a necessity. Large plants could achieve much better quality control, allowing tighter tolerances, higher quality concrete and better finishes than site-cast concrete.

Prestressed concrete technology began in Europe and, before the 1950s, was not widely used in the United States. The Pacific Northwest would soon become a leader in the American development of prestressed concrete thanks to a single individual—engineer Arthur Anderson (1910–1995). Anderson, a native of the Northwest who attended the University of Washington and MIT, was able to apply the frontier ingenuity and natural resources of the Pacific Northwest to the potential of prestressed concrete emerging a half a world away. While Anderson had no direct involvement with the US Science Pavilion, his essential role in making precast concrete a viable option in Seattle is undeniable.

Anderson worked on the landmark Pennsylvania Walnut Lane Bridge in 1950, establishing an instrumentation program for the prototype bridge. During this project, Anderson met the Belgian engineer Gustave Magnel (1889–1955).[13] Magnel was a pioneer in European prestressing, having developed the necessary techniques. The Walnut Lane Bridge was the beginning of a friendship between

Magnel and Anderson. But upon discussion with his brother Thomas back in Tacoma, Washington, Anderson decided to return to the Pacific Northwest and initiate their own prestressed concrete company.

Before doing so, the Anderson Brothers (and their father) took a three-week tour of Europe in October 1950, to see the most advanced prestressing operations in person. They visited France, Switzerland, Belgium, Sweden and England—including a visit with Magnel at his laboratory at the University of Ghent. Anderson described the lab as "the most advanced research center in the world," and Magnel was "very helpful and generous in sharing his knowledge."[14]

Even though he had observed operations throughout the world, Anderson knew that his own facility would have to be a combination of techniques used elsewhere and techniques developed on his own. The Andersons put their experience into place, investing over $200,000 in the construction of their own precasting plant with administrative building and prestressing beds. In 1954, Magnel visited the United States on his last lecture tour and came to Tacoma. Magnel was deeply impressed by the Andersons' operation—stating that his prestressing work was done with the utmost perfection.[15] In the years to come, the Andersons would make precast concrete a dominant part of the construction industry in the Pacific Northwest. Anderson and his firm cast a stadium grandstand in Tacoma (Cheney Stadium) and prestressed concrete beams for a downtown skyscraper (the Norton Building).[16] At the 1962 World's Fair, Anderson's firm would produce curved, prestressed concrete beams for the Fair's iconic monorail.[17]

In an effort to expand the knowledge and use of prestressed concrete in the region, Anderson began teaching a class in prestressed concrete class at University of Washington—open to both students and practicing engineers. At the time, few universities across the country were offering instruction in prestressing, and the course opened up a world of design possibilities to many. While Anderson did not have a specific role on the US Science Pavilion, his expertise was transmitted through two other individuals central to the Pavilion's success: precast concrete producer John L. Hutsell (1922–2010) and structural engineer Jack Christiansen (1929–2017). Both individuals took Anderson's course at the University of Washington and brought precast, prestressed concrete into their work.

Precast Concrete Production

The construction of the US Science Pavilion presented two pressing logistical questions. Who could cast the precast panels to Yamasaki's exacting specifications, and where? Thankfully, John Hutsell and the firm he worked for, Associated Sand and Gravel, in Everett, Washington, fit the part. Hutsell first studied chemical engineering at the University of Washington, but his studies were interrupted by the outbreak of World War II in 1939.[18] Returning to Seattle after the war, Hutsell began a career in concrete at Permanente Cement in Seattle. In 1950, Hutsell joined Associated Sand and Gravel in Everett, Washington—a concrete paving and asphalt firm, producing roadways and paved surfaces in the Puget Sound region. The firm had cast concrete utility pipes, built the grandstand for the Everett Memorial Stadium and was interested in expanding to other structural uses of precast concrete.[19] Hutsell became the Concrete Products Division Manager and helped grow the firm to one of the preeminent concrete firms in the country.

Hutsell engaged in the US Science Pavilion project very early on with the backing of his firm. Yamasaki's tight specifications for the texture and finish of the buildings required a specific formwork system and aggregate mix for an exposed-rock effect. Taking the initiative and seeing an opportunity, Hutsell offered his services prior to formally bidding on the project. Hutsell provided Yamasaki an array of different aggregate samples for the permanent glistening white finish of the building's walls to choose from and promised that he could deliver.

Hutsell recognized that forming the precast pieces for the Pavilion, both panels and long span beams, would require acute attention to detail and custom-built formwork. Once the aggregate was selected for the project, the design team decided to cast each panel with two layers of different mixes: a white facing mix and a gray structural backing mix. The facing concrete consisted of a white quartzite aggregate from Utah, crushed to a 95 percent minus 1/4-inch size, and a white cement. The backup mix was a high-strength structural concrete, containing both conventional reinforcing and prestressing cables. The two-mix approach had several complications relating to the bonding of dissimilar layers under high stress conditions. If not cast and stressed properly, the two layers would delaminate. To accommodate this, Hutsell placed an equal number of stressing strands in each layer and ensured they were equidistant from the center of the composite beam. The beams were then left to cure in a light- and temperature-controlled environment and turned on their side, so the different layers were side-by-side (not on top of each other). This process, combined with tight quality control of the two concrete mixes and attention to detail, produced high quality panels (Figure 9.4).[20]

Hutsell designed a glass-reinforced polyester laminate form, braced with a structural steel backup member for strength and precision. He also had custom cable-stressing equipment designed and built by his company engineering staff to meet Yamasaki's demands for tight tolerances and absolute thinness. The result of these specification was a new, state-of-the-art precast concrete production facility that could produce precast panels with variations no more than plus or minus 1/8th inch.[21]

The high cost of building a new facility could only be offset by repetitively using a single form to cast multiple panels. Associated Sand and Gravel could only find economy in the project through repetitive casting of identical panels, necessitating that the US Science Pavilion be composed of identical panels. Yamasaki visited the production plant in Everett on several occasions for consultation on the specifics of the precast production and to agree to minor design variations that would improve the economy of casting. In the end, the Pavilion buildings

FIGURE 9.4   Precast panels on site.

Source: Courtesy of Seattle Municipal Archive Digital Collection.

consisted of just three panel types—a bearing wall panel, overhead beams and end-wall (enclosing) panels (Figure 9.5). Hutsell found repetition in the casting of 364 bearing wall panels and 131 end wall panels.

Structural Engineer for the Pavilion

Though simple in their composition, each building had to resist all structural loads through very small structural members. Each panel was described as a thin-shell wall section, only 3 inches thick, between thickened ribs that served as the structural support for the overhead beams. This type of precision needed a tightly controlled production facility and specificity in its engineering design. Uncertainty of structural behavior would lead to a thickening of all the elements.

FIGURE 9.5   Precast beam and wall panels assembled.

Source: Courtesy of Seattle Municipal Archive Digital Collection.

In need of a good engineer to execute a technologically advanced building, Perry Johanson of NBBJ introduced Yamasaki to John Skilling and the engineers at Worthington Skilling Helle and Jackson. At Johanson's insistence, Yamasaki became familiar with Jack Christiansen's work in thin shell concrete and was particularly impressed by his use of prestressing to reduce the thickness of his shell structures.[22] Yamasaki agreed to hire Worthington Skilling Helle and Jackson as consulting engineers.

The curving ribs and columns embedded within each panel formed a delicate pattern that had to be bright, crisp and clean. They also had to be strong enough to carry the loads from the roof and floors of the buildings to the foundation. In order to achieve the desired thinness, Christiansen used Yamasaki's patterning as the beginning of his structure design by placing the reinforcing steel and prestressing tendons within Yamasaki's curving lines, leaving little flexibility to alter the

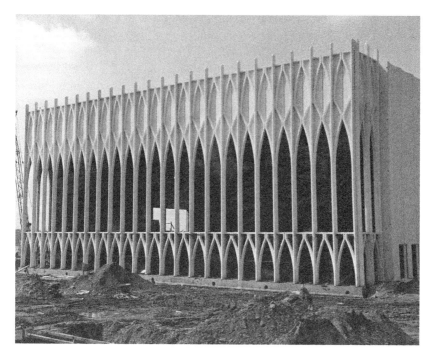

FIGURE 9.6   US Science Pavilion interior facade.

Source: Courtesy of Seattle Municipal Archive Digital Collection.

overall form. Christiansen's attention to detail facilitated Yamasaki's vision and insistence on thinness (Figure 9.6).

US Science Pavilion

With Hutsell and Christiansen providing the cutting-edge expertise, the project exceeded Yamasaki's expectations. In a statement titled "Design of the United Stated Science Exhibit, Seattle World's Fair" written after the Fair, Yamasaki made it clear that precast concrete was the only material option that could achieve the Pavilion's charge.[23] He stated:

> The precast quartz façade, we knew, would have sharpness of detail and elegance of appearance, and a permanent quality and richness which could not be gained in any other way.[24]

For Yamasaki, the designing for the future meant designing for permanence. It was one thing to create a building that looked good for the Fair, but something different to create a beautiful building that could persist. Precast concrete had an enduring quality that Yamasaki knew would last into the future. But precast concrete also served, on a conceptual basis, as a symbol of the future. Yamasaki stated:

> Moreover, in an important building such as the Federal Building for the World's Fair, we believed it was necessary to build with the most modern technological methods possible. This would then be totally in keeping with its purpose: That of the exhibition of scientific knowledge.[25]

FIGURE 9.7 Arches in the interior courtyard of the US Science Pavilion.

Source: Courtesy of Seattle Municipal Archive Digital Collection.

Yamasaki posited that the architecture required evidence of technological progress that came with advancement of scientific knowledge.

The buildings, towers and entrance platforms created a powerful composition, all executed in prestressed concrete (Figure 9.7). Architectural critique of the US Science Pavilion recognized the modern technological methods used to create the pavilion.[26] Yamasaki furthermore stated his approval in the *Seattle Times*. He said: "We're [Perry Johanson and Yamasaki] tickled to death. It's even more exciting that we thought it would be."[27] He continued: "For a science pavilion to show the progress of mankind, technologically we should be way ahead in construction. We think the building represents some real technological achievements, with prestressed, precast concrete techniques."[28] The US Science Pavilion won several awards, including an Award of Merit from the Precast/Prestressed Concrete Institute, described by the judges as a "virtuoso performance, showing what can be done with concrete."[29]

Conclusion

The US Science Pavilion, as an embodiment of the future, has lived up to its original charge. Not only was the pavilion celebrated during the fair, but it has continued to offer a science-oriented education and optimism as the Pacific Science Center. The buildings are substantial but welcoming. They are crisp and bright, but not cold. To be inside the Pavilion is to be supported on an outdoor platform, with precast concrete all around you and over your head. For the Pavilion design team, the future lay not in the introduction of something new but in the re-conceptualization of the familiar. Through design, technology, precision and vision, the Pavilion was able to offer a different concrete, a new concrete. And indeed, the use of precast concrete would expand greatly after the fair. As more places built precast facilities, precast (and prestressed) concrete became widespread in building and infrastructure projects. Prestressed concrete remains a standard for bridge construction, and precast panels are often used for high-quality textures, as it is now part of the standard tool kit of architects and engineers around the country. Precast concrete at the US Science Pavilion, phenomenal in its structural and material behavior, projected the idea that the future could be exciting, welcoming and permanent.

Notes

1. Paula Becker, Alan J. Stein, and History Link, *The Future Remembered: The 1962 Seattle World's Fair and Its Legacy* (Seattle, Washington, DC: Seattle Center Foundation, 2011), 11–15.
2. *Century 21 Exposition,* Century 21 Exposition, Inc., California Institute of Technology, 1959. 10.
3. Letter from Secretary Mueller to Mr. Philip Evans, Department of Commerce. December 22, 1959. Box 44, RG43, Records of the US Commission for the US Science Exhibit at the Seattle World's Fair, 1956–1963, National Archives at Seattle 1.

4. "Hall of Science Architect Named," *Seattle Times*, January 22, 1960, 1.
5. For an in depth look at Yamasaki's life and career, see Dale Allen Gyure and Minoru Yamasaki, *Minoru Yamasaki: Humanist Architecture for a Modernist World* (New Haven, CT and London: Yale University Press, 2017).
6. "Hall of Science Architect Has Impressive Record," *Seattle Times*, January 23, 1960, 3.
7. Letter from Minoru Yamasaki to Mr. Philip Evans, Department of Commerce. February 12, 1960. Box 44, RG43, Records of the US Commission for the US Science Exhibit at the Seattle World's Fair, 1956–1963, National Archives at Seattle 1.
8. Ibid.
9. Yamasaki stated: "Since this is a building which has very heavy occupancy and probably quite inflammable material within the building (the displays), it is of paramount importance that we have a fire-proof building. When a structure such as this necessitates fireproofing, then it is generally as economical in concrete as it would be in wood fireproofed or light steel fireproofed. We would recommend a concrete structure for a permanent building thus the structure would be essentially the same in cost whether it is temporary or permanent.

   As we see it, the elements which would be different in a permanent building would be the exterior wall, which might be cement plaster or stucco in a temporary building, and a more permanent finish such as aggregate surface concrete or brick in the permanent one, and the surface of the paving of the various terraces." Letter from Minoru Yamasaki to Mr. Philip Evans. Department of Commerce. February 15, 1960 in Box 44, RG43, Records of the US Commission for the US Science Exhibit at the Seattle World's Fair, 1956–1963, National Archives at Seattle 1.
10. "UW Alumnus of the Year Was 'Regular Fellow'," *Seattle Times,* June 6, 1960.
11. "Soaring Ribbed Vaults to Dominate Yamasaki's Design for Seattle Fair," *Architectural Record* 128, no. 8 (August 1960): 147–48.
12. Yamasaki, Minoru, *A Life in Architecture* (New York: Weatherhill, 1979), 70.
13. Charles C. Zollman, "Magnel's Impact on the Advent of Prestressed Concrete," *Reflections on the Beginnings of Prestressed Concrete in America,* Prestressed Concrete Institute, Chicago, IL. Reprinted from the copyrighted *Journal of the Prestressed Concrete Institute* 23–25 (May/June 1978–May/June 1980).
14. Arthur R. Anderson, "An Adventure in Prestressed Concrete," in *Reflections on the Beginnings of Prestressed Concrete in America* (Chicago, IL: Prestressed Concrete Institute, 1981), 189–237. Reprinted from the copyrighted *Journal of the Prestressed Concrete Institute*, 23–25 (May/June 1978–May/June 1980).
15. Tom Watson, "Remembering Art Anderson and Gustave Magnel," *PCI Journal* 49, no. 5 (2004): 115.
16. Gale Hemmann, "Concrete Technology: A Rock-Solid Piece of Tacoma's History," *South Sound Talk*, October 29, 2018. Tyler Sprague, "Norton Building," in *SAH Archipedia*, eds. Gabrielle Esperdy and Karen Kingsley (Seattle, Washington, Charlottesville: UVaP, 2012), accessed February 28, 2019, http://saharchipedia.org/buildings/WA010330075.
17. "90-foot Beams for Monorail Soon to be Delivered," *Seattle Times*, August 23, 1961, 1.
18. "John L. Hutsell—Obituary," *Seattle Times*, October 10, 2010.
19. "Contracts," *Seattle Times*, March 7, 1956.
20. John L. Hutsell, "Fabrication of Science Pavilion Wall Panels," in Harlan Edwards, ed., "Concrete Construction for the Century 21 Exposition," *ACI Journal Proceedings* 60, no. 6 (June 1963): 673–718.
21. Ibid.
22. For more on Christiansen's work, see Tyler S. Sprague, *Sculpture on a Grand Scale: Jack Christiansen's Thin Shell Modernism* (Seattle: University of Washington, 2019).
23. Minoru Yamasaki, "The Design of the United States Science Exhibit Seattle World's Fair," June 25, 1962 in Box 44, RG43, Records of the US Commission for the US Science Exhibit at the Seattle World's Fair, 1956–1963, National Archives at Seattle 1.

24. Ibid., 2.
25. Ibid., 3.
26. "10. Architectural Critique of the Pavilion," in *U.S. Science Exhibit—Seattle World's Fair Final Report* (Washington, DC: US Department of Commerce, Library of Congress Catalogue Card No. 63–60022, 1963), 51–56.
27. "Architects 'Tickled to Death' With Their Science Pavilion," *Seattle Times*, April 5, 1962, 12.
28. Ibid.
29. "2 Seattle Buildings Win Awards," *Seattle Times*, September 22, 1962.

10

THE CONCRETE FACADES OF PAUL RUDOLPH'S CHRISTIAN SCIENCE BUILDING, 1965–1986

Scott Murray

In March 1986, cranes with wrecking balls approached the Christian Science Building in Champaign, Illinois, on the University of Illinois campus. The arduous task of leveling a reinforced concrete building had begun (Figure 10.1). When the building opened just two decades earlier in 1965, its designer, Paul Rudolph, was among the most prominent of American architects. The building was initially praised in the architectural press and was considered emblematic of the growing global influence of Brutalist architecture at the time.[1] *Architectural Record* featured the building on the cover of its February 1967 issue, describing it as a demonstration of Rudolph's virtuosity.[2] Despite its earlier acclaim, after only two decades in use the building was abandoned, sold to a real estate developer and demolished in 1986, to be replaced by an architecturally unremarkable apartment complex.[3] Although still active professionally, Rudolph's career was also then in decline. He was consigned to the margins of his profession, and aside from some local protests and newspaper coverage, the destruction of the building went largely unnoticed on the architectural scene. Today, despite renewed appreciation of Paul Rudolph and his architecture, the Christian Science Building remains largely forgotten, and its premature destruction marks a significant loss of modern architectural heritage.[4]

The Christian Science Building was a relatively small yet spatially complex headquarters commissioned by and built for a religious student organization.[5] Composed as a series of interlocking rectilinear volumes and planes, the design was in many ways emblematic of Rudolph's work at the height of his career and shared several characteristics with his best known work, the Art and Architecture Building (A&A) at Yale University (1963). The Christian Science Building, like the A&A, was defined by two predominant characteristics: (1) an interior experience of spatial variation and contrast resulting from multiple floor levels, changes

FIGURE 10.1   *Left*: Paul Rudolph, Christian Science Building, Champaign, Illinois, 1965, main entrance at east facade. *Right*: demolition of the Christian Science Building, 1986.

Source: *Left*: Photograph HB-29509-Q, Chicago History Museum, Hedrich-Blessing Collection. *Right*: Photograph by Myra Kaha.

in ceiling height and dramatic daylighting effects; and (2) an exterior expression of solidity and handcraft conveyed through walls of cast-in-place concrete with a roughly textured, bush-hammered surface treatment, resulting in a distinctive pattern of vertical corrugations and light/shadow effects (Figure 10.2).

In the United States, demolition of a 20-year-old building is unusual, especially one designed by an accomplished architect, commissioned by a religious organization and affiliated with a major research university. Such institutions normally have a long-term outlook for buildings and are more likely to value architecture as cultural heritage than commercial property owners. Why, then, was the Christian Science Building destroyed? Gradual decline in demand for the original use of the building was a primary factor. However, many buildings facing this condition are deemed adaptable to new uses and find new owners who modify them to meet their needs. In this light, obsolescence, by itself, is not an adequate explanation. Another major contributing factor in the decision to destroy the building was the reportedly poor thermal performance of its concrete facades, which can in turn be traced back to a fateful decision made during the design process. However, like obsolescence, poor technical performance alone is not adequate to explain the destruction of such a recent building, considering that many older buildings are poorly insulated and can be improved through

FIGURE 10.2 Christian Science Building, east facade.

Source: Photograph HB-29509-A, Chicago History Museum, Hedrich-Blessing Collection.

retrofitting. The perceived cultural value of the building—and the evolution of such perception over time—must also be acknowledged as an influential factor in determining its fate.

A building facade always performs simultaneously in two realms: the technical and the cultural. While each is not exclusive of the other, acknowledgment of this duality is useful for comprehending how a facade functions and how it communicates ideas. In both realms, the design of a facade conveys embedded values and priorities, encompassing both quantitative and qualitative performance. A building envelope must respond to local climate and maintain desired interior environmental conditions; its thermal, hygric, daylighting and energy performance depend upon the facade's form, its materials and their assembly.[6] As the outward, projective and primary character-giving image of a building, the facade also expresses a building's cultural identity through its form, materiality and detailing. The communicative or expressive potential of a facade thus transcends technique but is not separate from it, and one may "read" a facade to gain an understanding of the priorities, both technical and cultural, of its designers, owners and/or constituents.

The concrete facade of Rudolph's short-lived Christian Science Building clearly communicates an aesthetic agenda, which was partly a rebuke of the predominant International Style "glass-box" architecture of the 1950s. But, in hindsight, it can also be read as a narrative about the building's cultural value, its perceived failures, its abandonment and its ultimate destruction. This understanding may have resonance beyond this single building, as it implicates the technical performance of facade design with a building's longevity and resilience in the face of economic and social change. From a cultural standpoint, these considerations are central to efforts to preserve modern architecture of the twentieth century and serve as a cautionary tale for the pursuit of architecture that is intended to be long-lived and sustainable in the twenty-first century.

Context and Building

Paul Rudolph (1918–97) established his practice in Sarasota, Florida, in the late 1940s, after completing his architectural education at Alabama Polytechnic Institute and Harvard University's Graduate School of Design (where he studied with Walter Gropius and Marcel Breuer).[7] During the 1950s in Sarasota, he designed a series of influential, internationally recognized houses during the 1950s, including the Healy House (1950), the Walker Guest House (1953) and the Hiss "Umbrella" House (1954). In 1958, Rudolph was appointed Chairman of the Department of Architecture at Yale University—a position he held for seven years. He was soon engaged in commissions for much larger institutional projects, including the Blue Cross and Blue Shield Building (1960) in Boston, the Temple Street Parking Garage (1963) and Yale Art and Architecture Building (1963) in New Haven, Connecticut. Rudolph's work at this time was characterized by an expression of monumentality bolstered by the predominant use of concrete. His design aesthetic aligned with the rising trend toward Brutalism in modern architecture.[8] With the move to Yale, Rudolph had achieved even greater professional recognition and acclaim and continued a prolific rate of production, completing more than 40 built and 30 unbuilt projects in the 1960s.[9] A profile of Rudolph in *Time* magazine in 1960 called him "the most exciting new arrival" on the architectural scene.[10] Robert A. M. Stern described Rudolph as "the greatest talent of his generation of American architects" who achieved a "meteoric rise to the top of the profession in the years following World War II."[11]

In 1962, Rudolph accepted a commission to design the new Christian Science Building in Champaign, Illinois.[12] He was chosen for the project by Nancy Houston, Director of the Building Trustees for the Champaign-based Christian Science student organization.[13] Project collaborators included a local Architect of Record, Delbert R. Smith, structural engineer Herman D. J. Spiegel and mechanical engineers Van Elm Heywood and Shadford.[14] Construction began in October 1964, led by the Felmley-Dickerson Company as general contractor, and the building opened in October 1965. Construction costs totaled $325,000, funded by private donations.[15] The building's vertical structure consisted of cast-in-place concrete foundations supporting load-bearing walls of cast-in-place concrete, while floors and roofs were framed with exposed wood beams and joists supporting plywood decking. At 5,630 square feet, the building represented a link between the more intimate scale of Rudolph's early residential work and the monumental scale of the institutional projects his office was increasingly pursuing. The main interior spaces included a lobby, a large meeting room with a lectern and seating for 75, smaller study rooms, a kitchen, a lounge, restrooms and a private studio apartment for a staff member. Ostensibly a two-story building, the interior actually contained seven different floor levels, with adjacent spaces often placed a few steps up or down from one another. Second-level spaces and interior bridges overlooked double-height spaces below. Ceilings ranged from just over 7 feet up to 42 feet in height. Many of the interior furnishings and light fixtures were designed by Rudolph and painted in vivid colors. From the exterior, the building appeared as an assemblage of cubic volumes and planes formed by walls of vertically corrugated concrete (Figure 10.3). This composition was punctuated by three towers, each topped by clerestory windows that transmitted daylight into the primary interior spaces.

In a 1968 letter to architecture students at the University of Illinois, Rudolph described his objectives for the Christian Science Building, highlighting the importance of the idea of scale to the design of the facades. Because the site was located at a point of transition on the campus, between what he termed the "huge scale" of university buildings to the east and the "more delicate scale" of single-family houses to the west, Rudolph wrote that he intended the Christian Science Building to appear "basically scaleless" so that "it can be read as more than one size."[16] To enable such ambiguity, facade elements that typically indicate scale, such as doors and windows, were deeply recessed or otherwise disguised. Traditional windows were not used; instead, glass was treated more abstractly as a transparent planar surface that periodically replaces concrete, creating voids in the solid mass. These openings introduced daylight to interior spaces and often revealed the thickness of the concrete walls. Rudolph wrote that the building "attempts to turn the corner, reads as a series of planes rather than a linear composition . . . subjugates clarity of structure for spatial variety, and depends on the introduction of lighting and color for its effect" (Figures 10.4 and 10.5).[17] The interior spatial variety, dramatic use of daylighting and careful attention to material details are characteristics that Lydia Soo and Robert Ousterhout, architectural historians at the University of Illinois, wrote "gave the interior a moving and meaningful character and the building a sense of place in the community"; in total effect, they observed, "it was a building that fulfilled practical needs and spiritual needs as well."[18]

The most distinctive aspect of the building, in both visual and tactile terms, was the roughly textured, bush-hammered surface treatment of its load-bearing reinforced concrete walls, exposed throughout the building on both the exterior and interior sides. This effect made the Christian Science Building unlike anything previously built in Champaign and, for those familiar with his work, immediately recognizable as a Rudolph design. As a concrete trade journal noted in

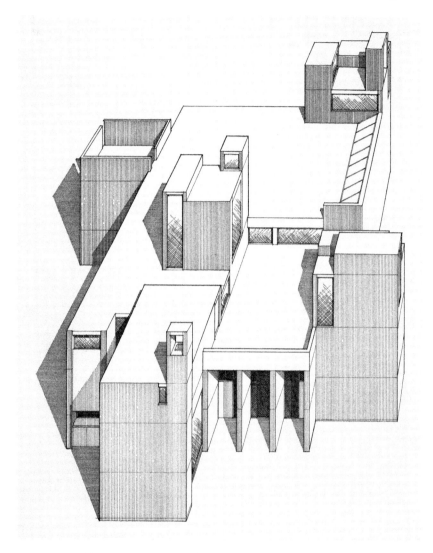

FIGURE 10.3   Christian Science Building, axonometric drawing by the office of Paul
Rudolph.

Source: Paul Rudolph Archive, Library of Congress, Prints & Photographs Division, PMR-2120.

1964, "[Rudolph's] corduroy-like, textured surfaces have become something of a
trademark."[19] To achieve this effect, the concrete walls were cast on-site in corrugated formwork consisting of plywood sheets to which narrow strips of wood
were nailed in a vertical pattern.[20] After curing, the formwork was removed,
and the sharp edges of the corrugations were bush-hammered by hand, a labor-
intensive process resulting in irregularly angled facets and exposed aggregate. This

FIGURE 10.4   Christian Science Building, meeting room interior.

Source: Photograph HB-29509-M, Chicago History Museum, Hedrich-Blessing Collection.

texture created delicate patterns of light and shadow, reflected light at different angles and further contributed to the ambiguity of scale.

Rudolph had first used this type of vertically striated bush-hammered concrete two years earlier at the Yale A&A Building and would continue to employ it in later projects as well. He was drawn to this treatment of concrete because, as he later wrote,

> it broke down the scale of the walls and caught the light in many different ways. . . . Light was fractured in a thousand ways and the sense of depth was

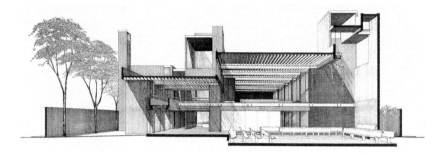

FIGURE 10.5   Christian Science Building, section-perspective drawing by the office of Paul Rudolph.

Source: Paul Rudolph Archive, Library of Congress, Prints & Photographs Division, PMR-2120.

increased. As the light changed the walls seemingly quivered, dematerialized, took on additional solidity.[21]

Beyond the fulfilment of their functional purpose to enclose space and to structurally resist loads, such facade effects were, in Rudolph's view, an important expression of cultural value. "Decoration adds meaning," Rudolph said, and "the best decoration has always grown in an integral way with the structure and reduced or heightened the scale."[22] With the integral decoration and handcrafted irregularity of the walls' corrugations and its emphasis on experiential qualities, Rudolph viewed the facade design as contributing to his ongoing efforts to "humanize" modernism and to challenge the dominance of the machine aesthetic in International Style architecture, which he believed had become monotonous, too reliant on an objective functionalist approach and devoid of human expression.[23] "I want buildings to move people," Rudolph said, and "mere functionalism is never enough."[24] However, eventually the Christian Science Building's capacity to move people and even to function would be called into question, with drastic results.

Crisis

By the mid-1980s, the Christian Science Building faced a crisis of ownership. Membership in the organization had significantly declined, from more than 100 students at its peak to fewer than ten. The remaining members found it increasingly difficult to fund the maintenance and operation of their now oversized building.[25] The high cost of heating the building during winter, in particular, was cited as an exorbitant problem and eventually forced the group to close the building during the winter months.[26] Although Rudolph had called the Christian Science Building "one of the more successful of my buildings,"[27] its owners no longer saw it that way. In 1986, the building was sold for $250,000 (less than the cost of construction 20 years earlier) to Gloria Dauten, a local real estate developer who planned to demolish it and build a six-story $2.5 million apartment complex on its prominent corner site.[28] Facing public protests from preservationists as well as students and faculty at the university's School of Architecture, Dauten offered to sell the property to the university. But the university's administration declined, saying that they had no use for the building in its current configuration and that if they modified it to suit their needs, "we would be the architectural villains."[29] Dauten also claimed to have contacted Rudolph twice for his opinion, reporting that "he felt if the building no longer served the need for which it was created, demolish it."[30]

The title of a front-page article in the local *News-Gazette* on February 2, 1986, succinctly summarized the building's impending plight: "With Demolition Approaching, Landmark Doesn't Have a Prayer."[31] The reporter noted that

> inside the thick concrete walls, it feels colder than the winter outside . . . the furniture upholstering and carpet are frayed and faded, as is the glory that used to make [the building] a popular feature subject in architectural trade magazines.[32]

In an interview conducted at Rudolph's New York office in that same month— just weeks before the building was demolished—Robert Bruegmann asked the architect for his thoughts on the Champaign building, and Rudolph recalled that "at that time I liked the building, and I still like it, which I can't say about all my buildings. I will be unhappy if its demolished, but I'm not in control of that."[33] Despite the protestations of preservationists, the demolition was carried out in March of 1986, beginning on the first day of spring break, perhaps strategically, when many faculty and students were absent from campus. Initially scheduled for two days of demolition, it ultimately took two weeks to take down the robust structure.[34] The destruction of the building echoed the decline of Rudolph's professional influence. By the 1980s, a Postmodernist critique of Brutalism— and Rudolph's work specifically—had taken hold. Rudolph had virtually disappeared, like the Champaign building, from architectural discourse.[35]

A Tale of Two Walls

In Bruegmann's 1986 interview, Rudolph reveals an interesting fact about the design of the Christian Science Building's facades: the architect had originally proposed building them with concrete block, not cast-in-place concrete. At the time of the design, Rudolph had recently begun collaborating with Plasticrete, a masonry manufacturer in New Haven, to develop a series of customizable, factory-produced "fluted" blocks that, when assembled into a wall, would resemble the vertical corrugations of the A&A Building.[36] Rudolph believed that using blocks

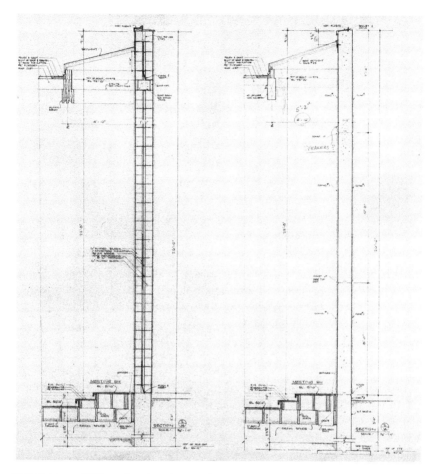

FIGURE 10.6   Wall section drawings in the construction documents set by the office of Paul Rudolph showing the original design of cavity wall with custom-fluted concrete block and insulation (*left*) and revised design of monolithic cast-in-place reinforced concrete without insulation (*right*).

Source: Paul Rudolph Archive, Library of Congress, Prints & Photographs Division, PMR-2120.

in the Champaign project would be less expensive than cast-in-place concrete, but the builders disagreed with him, saying that the block walls would be more costly to build. Although the builders' position runs counter to the widely accepted notion that prefabricated concrete-block construction is categorically less expensive than site-cast concrete, Rudolph had not yet completed any projects with the fluted blocks and said he was unable to prove its cost savings. In the end, the Christian Science Building's walls were changed to cast-in-place construction.[37]

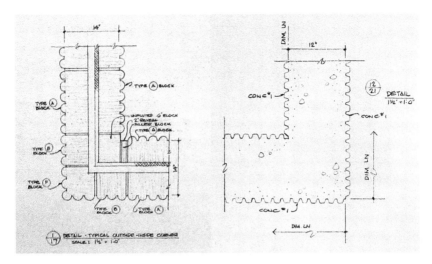

FIGURE 10.7   Corner plan details by the office of Paul Rudolph show the original design of cavity wall with custom-fluted concrete block and insulation (*left*) and revised design of monolithic cast-in-place reinforced concrete without insulation (*right*).

Source: Paul Rudolph Archive, Library of Congress, Prints & Photographs Division, PMR-2120.

Two sets of construction drawings produced by Rudolph's office confirm and document the change in wall design that Rudolph had recalled in the 1986 interview.[38] One set of drawings, with an issue date of June 15, 1964, shows the exterior walls consisting of two wythes of custom-formed concrete blocks with an interstitial air cavity and a layer of foamboard insulation (14 inches total wall thickness), while in a subsequent revision, dated November 11, 1964, the walls were all revised to monolithic cast-in-place reinforced concrete in a single layer (12 inches total wall thickness) with no cavity or insulation (Figure 10.6). On some drawings, one can see evidence of the erasure of the multiple layers of the earlier cavity walls. In the pursuit of lower costs, there was apparently little value placed on the thermal performance of the wall; along with this design change came the elimination of the thermal insulation and air cavity of the original assembly.

The different wall types are clearly evident in each drawing set's floor plans, building sections, wall sections and details (Figure 10.7). In either case, how-ever, the same corrugated effect would be achieved. The drawings show that Rudolph's office had designed a set of custom-shaped corrugated blocks with convex ridges running vertically, including typical field blocks and special corner blocks that would produce the desired vertical striations like those achieved by the bush-hammered cast-in-place walls (Figure 10.8).[39]

The fluted blocks would be laid in a stacked bond pattern (instead of running bond), with the vertical joints aligned and located in the recess between two

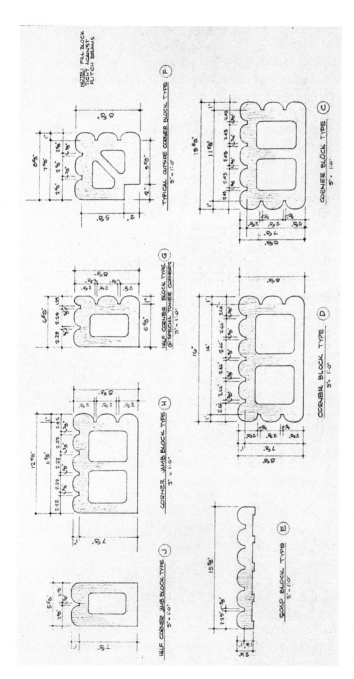

FIGURE 10.8   Concrete block types from initial drawing set by the office of Paul Rudolph; later revisions replaced block walls with cast-in-place concrete.

Source: Paul Rudolph Archive, Library of Congress, Prints & Photographs Division, PMR–2120.

flutes. The color of the mortar would match the concrete, thus minimizing the appearance of individual blocks and creating a visually continuous surface similar to a poured concrete wall.[40] Encapsulating the air cavity and insulation layer, the two layers of blocks would create corrugated concrete surfaces exposed both outside and in. In both sets of drawings, the building's other envelope components—its glass and roof assemblies—remained consistent. Single-pane glass was used for windows and skylights, and built-up roofing with gravel ballast covered 2 inches of rigid insulation over plywood decking.

Soon after completion of the Christian Science Building, Rudolph successfully used custom-textured, prefabricated concrete blocks in other projects, as he had originally proposed for Champaign. At Crawford Manor, a 14-story apartment building in New Haven, opaque walls were formed with fluted blocks almost identical in design to those indicated in the earlier Christian Science Building drawing set. The Charles A. Dana Creative Arts Center at Colgate University used a split-rib block, a rougher version in which concrete blocks were cut in half, exposing the fractured internal ribs which, when stacked, formed vertical fins similar in appearance to the bush-hammered cast-in-place walls. Interestingly and somewhat confoundingly, the Colgate building's walls were originally designed to use cast-in-place concrete but in the final specifications were changed to concrete block construction as a cost-savings strategy—a complete reversal of the process and results at the Christian Science Building project.[41] Crawford Manor and the Creative Arts Center were both completed in 1966. Thus, the Christian Science Building was positioned chronologically between the cast-in-place concrete of the A&A Building, two years earlier, and the concrete block construction of these two projects that followed a year later.

Thermal Performance

In any construction project, decisions made during the design and construction phases may have long-term and sometimes unforeseen effects on the building's future operating costs. Given that the excessive cost of heating the Christian Science Building was cited as one of the major reasons for its eventual abandonment, the fundamental decision made by the project team about how to construct its facades takes on major significance in retrospect.

For context, it is useful to consider the thermal performance of the Christian Science Building's concrete walls *as built* in comparison with Rudolph's original plan to utilize cavity-wall construction. For perspective, a similar comparison can be made with today's energy code requirements. While it is unrealistic to expect that an architect working in the 1960s would design facades that meet today's standards for energy performance (and there was no energy code in place at the time), such comparisons can be instructive in assessing the building's purportedly poor thermal performance.

If it were built today, Rudolph's Christian Science Building would be subject to the requirements of a stringent energy code governing the design of the building envelope and environmental-control systems.[42] For Champaign's climate zone, the minimum required R-value for continuous insulation in a mass wall is R-11.4.[43] As built, the 12-inch thick cast-in-place concrete walls of the Christian Science Building contained no insulation material; the concrete alone would have achieved approximately R-1.3 to 1.5. The CMU cavity walls indicated in early issues of Rudolph's construction drawings indicate two wythes of custom-fluted 6-inch concrete blocks separated by a 1–3/4-inch air space and 1-inch thick foamboard insulation. The insulation alone would have achieved an R-value of approximately R-5, and the overall assembly would reach approximately R-8. It is also notable that Rudolph's construction drawings indicate 2 inches of rigid insulation located above the plywood roof deck, which would fall significantly below the code minimum of R-30 for roofs. While a singular focus on thermal resistance and isolation as a measure of energy performance has been rightly criticized,[44] it provides one lens through which to understand the relative performance of the Christian Science Building's concrete facade. Rudolph's original plan to use a CMU cavity wall would not meet today's energy-performance standards but would have performed much better in the Illinois climate than the as-built "value-engineered" wall did, theoretically achieving a nearly seven-fold increase in R-value.

What Might Have Been

The historian Daniel Abramson writes that architectural obsolescence is a particularly twentieth-century phenomenon:

> Before the twentieth century, building were subject to obsolescence, to be sure . . . but not until [then] did obsolescence come to be understood as a general condition of change in architecture and cities as a whole—a relentless, universal, impersonal process of devaluation and discard.[45]

In his book, *Preservation of Modern Architecture*, Theodore Prudon identifies perceived and/or actual obsolescence as a major challenge facing many works of modern architecture as they age.[46] He describes two types of obsolescence—functional and physical—noting that "functional obsolescence is understood in two ways: either the original use is no longer needed, or the building is deemed outdated due to the evolution of expectations."[47] Physical obsolescence, on the other hand, results when materials and building systems—often new and experimental in twentieth-century modern architecture—fail or are surpassed in their performance capabilities as more advanced technologies, standards and expectations are established. Both types of obsolescence can be contributing or even deciding factors in decisions to destroy works of modern architecture.

Prudon further explains that economic viability, which is often tied directly to a building's functional and physical performance, is particularly important for modern buildings. This is due to their often functionally determined form, which may be perceived as less conducive to adaptive re-use than traditional buildings.[48] Although Prudon did not write specifically about the Christian Science Building, his observations on obsolescence fit neatly with the building's history. It was subject to pressures of functional obsolescence (as student membership declined) that combined with its perceived lack of economic viability (occupying a site that could generate more profit with an apartment building) and the functional obsolescence of its envelope (consisting of uninsulated concrete walls and single-pane glass windows) to prematurely end its useful lifespan (Figure 10.9).

With obsolescence in mind, it is instructive to consider alternate trajectories that Rudolph's building in Champaign may have experienced, had different paths been taken regarding its facades. It is clear that Rudolph's original intent to use an insulated cavity wall would have provided more robust thermal insulation than the as-built cast-in-place wall. Would such a difference have impacted the building's longevity? It is of course impossible to know with certainty, but such an outcome seems at least possible considering that heating costs were cited as the building's fatal flaw. And if the building's thermal performance had been at a more acceptable level in the 1980s, perhaps other potential owners, such as the university or other student organizations, would have been more willing to take over ownership and operation of the building.

At the time of demolition, there were methods available to remedy inadequate insulation in the building envelope. Of the two main envelope components— roof and walls—the roof's lack of sufficient insulation could have most easily been remediated with additional layers of rigid foam insulation. The single-pane glazing used throughout the building could have been replaced with insulating glass, albeit at significant cost. Technically, it would have also been possible to consider various retrofit options for the concrete facades that would have improved the overall thermal performance of the envelope. An expensive option was to remove and reconstruct the concrete walls with a sandwiched layer of internal insulation. A more realistic alternative was to add a layer of insulation to the interior surfaces of existing concrete walls, covered with a new interior finish of concrete block, gypsum board or other material. However, while the latter approach would improve the R-value, it would have dramatically altered the architectural integrity of Rudolph's design, which placed such emphasis on the materiality of corrugated concrete forming nearly all of the interior *and* exterior surfaces. There is no evidence that such retrofit options were seriously considered by the owners of the building, which suggests that its fate was sealed by other factors, such as the loss of population in the student organization and the value of the land for other development purposes.

As expectations for energy performance continue to rise, strategies for mitigating the inherent deficiencies of decades-old building-envelope technologies

FIGURE 10.9   Demolition of the Christian Science Building, 1986. Graffiti reads: "Less is more, more or less. RIP 1965–1986."

Source: Photograph courtesy of Myra Kaha.

will continue to be a challenge for owners, architects and preservationists. A 2008 renovation of the A&A Building also sheds light on what might have been possible at the Christian Science Building. Along with a major new addition, the project involved renovating Rudolph's Yale building, restoring as much of the original design intent as possible (much of which had been incrementally compromised over decades of use), while also improving energy performance, life-safety systems and accessibility. Like the Christian Science Building, the A&A was built with monolithic, uninsulated concrete walls and single-pane glazing. The renovation did not modify the concrete walls, aside from repairing some surface damage, but did install new double-pane glass with a state-of-the-art low-e coating, thus improving thermal and solar-heat-gain performance in all windows and skylights. Working holistically, the project team utilized a trade-off approach, offsetting poor performance in the concrete walls with improved glazing, additional roof insulation and new high-solar-reflectance roofing, daylighting and occupancy sensors and new high-efficiency HVAC systems and controls.[49]

Resurgence

As popular perceptions shift, the work of architects like Paul Rudolph and other Brutalists has been recently reevaluated and has gained renewed appreciation.[50] Writing for the American Institute of Architects in 2015, Mike Singer notes in an article titled "Revisiting Paul Rudolph" that "today, a new generation of architects and design enthusiasts are paying homage to Rudolph in both word and deed."[51] After a string of demolitions of Rudolph projects in recent decades, including two of the three public schools he designed and several houses, preservationists have succeeded in saving and restoring some significant Rudolph projects. Among these are Sarasota High School (built 1959 and renovated in 2015) and the Jewett Arts Center at Wellesley College (built in 1958 and the subject of a comprehensive conservation plan funded by the Getty Foundation in 2015).

Theodore Prudon writes that a determination of the heritage value of a particular building results from a combination of professional and public perception and that, in the case of modern architecture, such perceptions often evolve over time and tend to evolve toward broader acceptance.[52] If the Christian Science Building had been able to avoid the problems associated with its technical performance, it may have benefitted from the current resurgence of interest in Paul Rudolph. In other words, if the technical performance had been bolstered early on, either through a more robust facade design as originally planned or through extensive renovation, there may have been time for its programmatic and cultural performance to be valued again.[53] Had it survived, the Christian Science Building and its facades would likely now be considered a landmark example of 1960s modernism and representative of the work of a historically important architect at the apex of his career.

Notes

1. Although often associated with Brutalism, Rudolph didn't use that term to describe his work. He admired several British proponents of the movement, including Alison and Peter Smithson, and appreciated its approach to architectural expression through

"rough, simple materials." See Timothy M. Rohan, *The Architecture of Paul Rudolph* (New Haven, CT: Yale University, 2014), 76.

2. "An Architecture Strongly Manipulated in Space and Scale," *Architectural Record* 141, no. 2 (February 1967): 137–42. Also see Nory Miller, "Rudolph's Rich Molding of Space at the U of I," *Inland Architect* 17, no. 9 (September 1973): 18–19.

3. Although the building was completely destroyed, some relics were saved prior to demolition: several pieces of Rudolph-designed furniture and light fixtures were retained by the university's School of Architecture, and the developer who purchased the building, Gloria Dauten, kept the massive front doors, made of brass and designed by artist Roger Majorowicz, later installing them as a decorative fireplace surround in her own home in Champaign.

4. Rudolph's work was the subject of a major exhibition and symposium at Yale University in 2008, which was followed by a monograph by Rohan, *The Architecture of Paul Rudolph*—marketed as "the first in-depth study of the architect, neglected since his postwar zenith" (from the book jacket)—and a collection of essays developed from the symposium in Timothy M. Rohan, ed., *Reassessing Rudolph* (New Haven, CT: Yale School of Architecture, 2017). Beyond a passing mention, the Christian Science Building is not discussed in either book.

5. For a comprehensive history of the building, see Scott Murray, "A Monumental Absence: Paul Rudolph's Christian Science Building, 1965 (demolished 1986)," in *Modernism and American Mid-20th Century Sacred Architecture*, ed. Anat Geva (London: Routledge, 2019), 132–52.

6. Among architects, there is not a singular definition of *facade*. Although it is sometimes understood in limited terms—signifying only the outermost surface of a building's wall, or even just the front surface of a building—here I take an expanded view of facade to include the entire wall assembly that forms the exterior envelope of the building, regardless of orientation or frontage, and which may be load-bearing (as in the case of a concrete or masonry wall) or non-load-bearing (as in a curtain wall) and includes the full extent of the wall's thickness, from outermost to innermost surface.

7. Rudolph worked in partnership with Sarasota architect Ralph Twitchell until 1952, when he established his solo practice.

8. Historian Marcus Whiffen writes: "In America exposed concrete left in its rough state—or sometimes, as in Paul Rudolph's Art and Architecture Building at Yale, artificially roughened—is common to a great many, if not most, of the buildings by which the adjective Brutalist comes to be applied." Whiffen, *American Architecture Since 1870: A Guide to the Styles* (Cambridge: MIT Press, 1969), 279.

9. See Project List in Rohan, *The Architecture of Paul Rudolph*, 250–60.

10. "Bright New Arrival," *Time*, February 1, 1960, 60.

11. Robert A. M. Stern, Foreword, in Paul Rudolph, *Writings on Architecture* (New Haven, CT: Yale University, 2008), 6.

12. In publications, the building has been identified by various names, including Christian Science Student Center and Christian Science Center. Rudolph's construction drawings use the project title "Christian Science Building," which is the name used throughout this chapter.

13. See "Architect Rudolph to Visit University," *The Daily Illini* (Champaign, IL), January 6, 1965, 9.

14. As noted on construction drawings in the Paul Marvin Rudolph Archive, Library of Congress, Washington, DC. For more information on the client/architect relationship and the commissioning process, see Murray, "A Monumental Absence," 134–36.

15. J. Philip Bloomer, "With Demolition Approaching, Landmark Doesn't Have a Prayer," *The News Gazette* (Champaign, IL), February 2, 1986, 2. The budget had previously been reported to be $150,000. See "Christian Science Foundation Plans New Building," *The Daily Illini* (Urbana-Champaign, IL), May 21, 1963, 10.

16. Paul Rudolph to Joseph Martinez and Thomas Heinz, December 17, 1968. Letter provided to the author by the Anthemios Chapter of Alpha Rho Chi at the University of Illinois at Urbana-Champaign.

17. Ibid.

18. Lydia M. Soo and Robert Ousterhout, "Has Functionalism Triumphed? The Destruction of Paul Rudolph's Christian Science Building," *Reflections* 4, no. 1 (Fall 1986): 43.

19. "Paul Rudolph: Artist in Concrete," *Concrete Products* (October 1964): 31.

20. For more on this technique, see Réjean Legault, "Paul Rudolph and the Shifting Semantics of Exposed Concrete," in *Reassessing Rudolph*, 81–83.

21. Paul Rudolph, "Enigmas of Architecture," in *100 by Paul Rudolph: 1946–74*, ed. Toshio Nakamura (Tokyo: A+U Publishing, 1977), 318.

22. Jeanne M. Davern, "Conversations with Paul Rudolph," *Architectural Record* 170 (March 1982): 91.

23. See Rohan, *The Architecture of Paul Rudolph*, 2–3, 33, and Soo and Ousterhout, "Has Functionalism Triumphed?" 42–43. Rudolph referred to glass curtain wall buildings as "goldfish bowls," writing in 1954 that "we build . . . too many goldfish bowls, too few caves." Paul Rudolph, "Changing Philosophy of Architecture," *Architectural Forum* 101 (July 1954): 120.

24. John W. Cook and Heinrich Klotz, *Conversations With Architects* (New York: Praeger, 1973), 96–97.

25. See Bloomer, "With Demolition Approaching, Landmark Doesn't Have a Prayer," 1.

26. As reported in Soo and Ousterhout, "Has Functionalism Triumphed?" 43.

27. Robert Bruegmann, "Interview With Paul Rudolph," February 28, 1986. Chicago Architects Aural History Project. The Art Institute of Chicago, 48–50, http://digital-libraries.saic.edu/cdm/ref/collection/caohp/id/9861.

28. Bloomer, "With Demolition Approaching, Landmark Doesn't Have a Prayer," 2.

29. Ibid., 1. Dauten also looked into the possibility of moving Rudolph's building to a different site but, unsurprisingly, found this to be unfeasible. Ibid., 2.

30. Bloomer, "With Demolition Approaching, Landmark Doesn't Have a Prayer," 2.

31. Ibid., 1.

32. Ibid.

33. Bruegmann, "Interview With Paul Rudolph." Rudolph had revisited the building recently, as he served as Plym Distinguished Professor of Architecture at the University of Illinois at Urbana-Champaign during the fall semester of 1983.

34. Soo and Ousterhout, "Has Functionalism Triumphed?" 40.

35. See Rohan, *The Architecture of Paul Rudolph*, 1, 177–79.

36. They were also sometimes called "corduroy" blocks. Rohan, *The Architecture of Paul Rudolph*, 141–43, and Legault, "Paul Rudolph and the Shifting Semantics of Exposed Concrete," 85–88.

37. Bruegmann, "Interview With Paul Rudolph."

38. Paul Marvin Rudolph Archive, Library of Congress, PMR-0664 and PMR-0686.

39. While visually similar, however, the two types of construction were conceptually quite different. As Legault writes: "With the precast block, the texture and ridges of Rudolph's original design at the A&A were now machine-made. Self-expression had been replaced by anonymous reproduction." Legault, "Paul Rudolph and the Shifting Semantics of Exposed Concrete," 88.

40. See Legault, "Paul Rudolph and the Shifting Semantics of Exposed Concrete," 85–88.

41. Ibid., 87.

42. If it were built today, the Christian Science Building would be required to comply with the Illinois Energy Conservation Code, which currently adopts a slightly modified version of the International Energy Conservation Code (IECC).

43. International Energy Conservation Code (IECC), Table C402.1.3.

44. For example, see Kiel Moe, *Insulating Modernism: Isolated and Non-Isolated Thermodynamics in Architecture* (Basel: Birkhauser, 2014), 10–53.

45. Daniel M. Abramson, *Obsolescence: An Architectural History* (Chicago, IL: University of Chicago, 2016), 5.

46. Theodore Prudon, *Preservation of Modern Architecture* (Hoboken: Wiley, 2008), 30–34.

47. Ibid., 30.

48. Ibid.

49. The project team included Gwathmey Siegel & Associates Architects, Hoffmann Architects and Atelier Ten. The renovation project was awarded a LEED Gold certification. See Russell Sanders, Benjamin Shepherd, Elizabeth Skowreonek and Alison Hoffman, "Sustainable Renovation of Yale University's Art + Architecture Building," *APT Bulletin: Journal of Preservation Technology* 42, no. 2–3 (2011): 39–45.

50. For example, despite a tumultuous history and a long estrangement of Rudolph from Yale, following its extensive 2008 renovation the A&A Building at Yale was renamed Rudolph Hall in honor of its architect.

51. Paul Singer, "Revisiting Paul Rudolph," American Institute of Architects, www.aia.org/articles/1736-revisiting-paul-rudolph:16.

52. See Prudon, *Preservation of Modern Architecture*, 26–27.

53. The revival of interest in Brutalism is not universal, of course. For example, writing in 2012, the critic Anthony M. Daniels refers to Brutalist buildings as "concrete monstrosities" and warns that "a single such building can ruin an entire townscape." Anthony M. Daniels, "Atrocities Should Be Eliminated," *NYT*, The Opinion Pages: Room for Debate, April 9, 2012, www.nytimes.com/roomfordebate/2012/04/08/are-some-buildings-too-ugly-to-survive/atrocities-should-be-eliminated.

11

BILL HAJJAR'S AIR-WALL

A Mid-Twentieth-Century Four-Sided Double-Skin Facade

Ute Poerschke and Mahyar Hadighi

Starting in the 1930s, architects, clients, the building industry, research institutions, scientific journals and non-scholarly magazines became increasingly interested in solar architecture. When in the late 1950s Abraham William Hajjar (1917–2000), known to his friends as Bill Hajjar, started his unique project on double-skin facades (DSFs) to advance solar heating of buildings, he could ground his work on 20 years of research and design experience in solar architecture. His research and design activities, undertaken with the support of Pittsburgh Plate Glass, provide insights into the manifold endeavors that intertwined the development of solar architecture and glass technology.

Bill Hajjar is regionally well regarded for his postwar single-family houses built in the area of State College, PA.[1] He is also remembered as an enthusiastic teacher during his time at Penn State between 1946 and the mid-1960s.[2] Comparatively important and overlooked, however, are his ambitious research endeavors. His main research focus was on four-sided DSFs, but he also worked on nuclear shelters. He developed a core-and-shell principle that was often grounded on a square floorplan and he was very successful combining this design principle with his research foci. It is this relationship of design and research that warrants reassessing his accomplishments.[3]

Bill Hajjar: Toward Solar Modern Architecture

Bill Hajjar was born on February 11, 1917, in Lawrence, MA, into an immigrant Lebanese family. The youngest of eight children, Hajjar decided against a career in the family's grocery store. In 1936 he enrolled at the Carnegie Institute of Technology (now Carnegie Mellon University) and graduated with a professional degree in architecture in 1940. He then pursued graduate studies in architecture at the Massachusetts Institute of Technology (MIT), receiving his Master of Architecture degree in 1941. After teaching in the Department of Architecture at the State College of Washington and serving in the military during World War II, he moved to State College, PA, in 1946, where he joined the Department of Architecture at the Pennsylvania State College (renamed Pennsylvania State University in 1953).

While Hajjar was at Carnegie Tech, the school's pedagogical philosophy, like most programs in the United States, was dominated by the Beaux-Arts approach. Yet, modern ideas of design found their way. For example, Walter Gropius, founder of the Bauhaus School, delivered a lecture on March 11, 1938, which provided the first opportunity for Hajjar to interact with one of the pioneering masters of modern architecture. Similarly, at MIT, Hajjar became familiar with modernism through advocates of modern architecture, including his advisor Lawrence B. Anderson. Anderson was one of a number of faculty members who were advocating for modern architecture at MIT during the 1930s. In 1939, with his office partner and fellow faculty member Herbert L. Beckwith, Anderson completed the MIT Alumni Pool Building, which was one of the first modernist buildings on an American campus. The building included a southeast-facing, partially double-layered glass facade that could have encouraged Hajjar to become interested in solar and glass architecture early in his career.[4] In addition, Anderson advocated for a new system of reviewing students' work, which relied on including assessments from critics outside the institute. He often invited Gropius and Marcel Breuer, who were teaching at Harvard's Graduate School of Design and practicing together, to participate in such reviews, thus providing his students, including Hajjar, direct exposure to modernist thinking in the late 1930s and early 1940s. After World War II, Anderson became very active in solar projects funded by the MIT Solar Energy Fund. From 1947 to 1948 he worked on an experimental building that became widely known as the MIT Solar House III. He was also responsible for the MIT Solar House IV, built in 1958. While these endeavors in solar architecture occurred after Hajjar's graduation from MIT, it is unlikely that he could have ignored their discussions in *Progressive Architecture*, *Architectural Forum* and other publications.[5]

Soon after gaining tenure at Penn State, Hajjar ramped up his practice. Between 1952 and 1963, he built single-family houses along with institutional and apartment buildings in and around State College, many of them with his office partner Harlan J. Wall. Later in the 1950s, he started his research career. His approach to solar architecture, with a focus on double-skin facades (DSF), was completely different to the research at his alma mater and elsewhere.

The Air-Wall Concept

It is not exactly clear when and how Hajjar became interested in DSFs. However, once started, his research almost immediately gained large-scale industry support from the Pittsburgh Plate Glass company and aroused attention from architects and the public. As early as 1958, Hajjar started with a design of a 96-foot square, seven-story office building surrounded by a DSF, which was to be 4 feet, 6 inches wide on axis, serving as "a blanket of circulated air"[6] (Figure 11.1).

The wall section that he presented to Pittsburgh Plate Glass in 1959 pushed the limit of existing facade technology and aesthetics alike (Figure 11.2). It integrated electrical lighting and a "radiant curtain" in the DSF, the latter to absorb heat throughout the year and to electrically heat the building in the cold seasons. The rendering showed no openings in the floors of the DSF, suggesting that air was anticipated to circulate horizontally. A floor plan revealed that vertical glass chimneys were positioned at each of the building's corners, which connected the DSF floors. These chimneys were surrounded by glass louvers to allow outside air

FIGURE 11.1   A. William Hajjar, 1958–59 design of an office building surrounded by DSFs.

Source: Special Collections Library, Pennsylvania State University. Courtesy of Mark Hajjar.

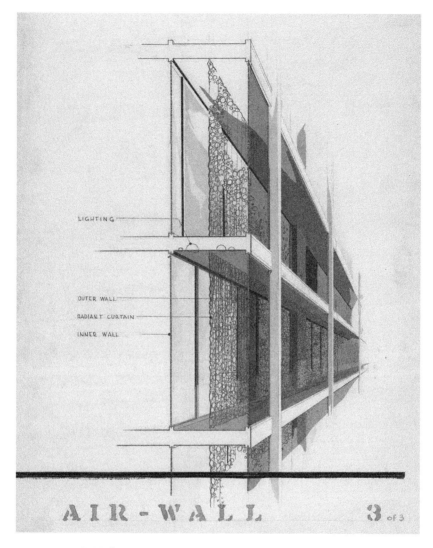

LIGHTING

OUTER WALL
RADIANT CURTAIN
INNER WALL

AIR-WALL   3 of 3

FIGURE 11.2 A. William Hajjar, 1958–59, facade section of DSF.

Source: Special Collections Library, Pennsylvania State University. Courtesy of Mark Hajjar.

to enter at one corner, flow horizontally in each story's DSF and exhaust through the glass chimney at the opposite corner.

The DSF and its added elements constituted a system for solar and electrical heating, as well as for daylighting and electrical lighting. The DSF floors and curtains were also intended to provide shading for the inner rooms and reduce heat build-up in summer, while mechanical cooling and air conditioning were not mentioned. A more technical drawing (Figure 11.3) reveals more details. The inner layer is composed of 1/4-inch single plate glass set in frames between

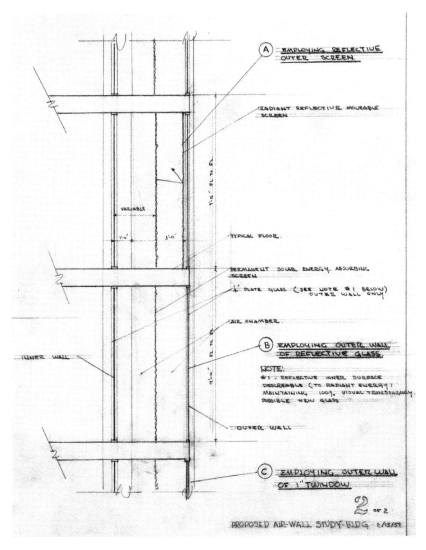

FIGURE 11.3 Air-Wall, facade section, including three alternatives for the outer glass layer.

Source: Special Collections Library, Pennsylvania State University. Courtesy of Mark Hajjar.

structural columns (no materiality of primary structure is mentioned). The outer layer of glazing, again set in frames, rests on an extension of the floor slab. Three different alternatives for the outer layer are proposed. The top floor employs 1/4-inch single-glazing and a "radiant reflective moveable screen" directly behind the glass, with the reflective surface toward the inside to trap "radiant energy." The floor below employs a 1/4-inch single "reflective glass" for the same

intention of trapping heat. The third option, shown at the bottom of the drawing, employs a 1-inch "Twindow," a double-paned insulating glass product that
PPG had launched in 1945. Halfway between the inner and outer glass layers,
Hajjar inserted a "permanent solar energy absorbing screen," for which he noted
in the floor plan drawing that it could additionally work as an "electrical resistive
heating element."

Hajjar anticipated manifold advantages for his facade system. A central goal
was to reduce air infiltration into indoor spaces. In the 1950s and 1960s, single-
pane glazing was still the common product for windows, and the additional glass
layer was intended "to relieve the inner wall from the complex weatherproofing
now so necessary."[7] Hajjar also thought to create an additional insulation layer
for winter and summer conditions and to protect the sunshades. Moreover, his
vision went beyond providing protection. In winter, he sought to have the sun
heat air in the southern double-skin cavity, which would then, by convection,
horizontally move towards the colder northern cavity with the effect of rendering consistent temperature conditions around the entire building. To make it an
active "built-in climate controller," he envisioned for the radiant curtain that in
"the cold seasons you simply add more warm air to the system and warm your
entire house without a complex system of ducts and vents."[8] The overarching
aspiration was to create an ideally tempered layer surrounding the interior space
in order to make heating or cooling within the rooms obsolete.

With the benefit of a design in hand, Hajjar convinced Pittsburgh Plate Glass
to fund an "Air-Wall Test Building" that was, relative to the initial design, modest in size but substantial for a test facility. In August 1959, ground was broken
for a four-story structure with an exterior dimension of 26 × 26 feet in plan.
Oriented to the cardinal directions, it included a ground floor with testing equipment and two upper floors surrounded by a DSF consisting of two layers of
1/4-inch single glazing, set three feet apart. An additional top floor, without a
DSF, was constructed to facilitate measurements for comparison to a conventional
facade. The DSF differed from the initial larger building design in significant
ways. The corner air chambers were abandoned and operable vents at the bottom, top and between the two DSF stories were inserted at all four building sides
(Figure 11.4).[9] This new set-up allowed an increased air mixing within the entire
double-story air cavity, while the study of separate stories was still possible when
vents were closed. The research team, now including Penn State's architectural
engineering professor, Vincent L. Pass, and mechanical engineering professor,
Everett R. McLaughlin, was eager to start its research and began measurements
even before the building was fully completed in mid-1960.

First measurements were conducted on March 12, 1960, and documented as
Results of Qualitative Exploratory Tests. Charts with exterior, interior room and
DSF cavity air temperatures were arranged in a diagram that mimicked the floor
plan (Figure 11.5 *left*) to allow easy comparison. The recorded measurements were
surprising. While the outdoor air temperature peaked at 39°F around 3:00 p.m.,

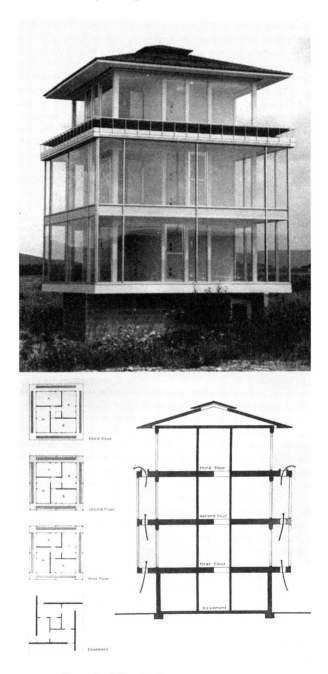

FIGURE 11.4   Air-Wall Test Building 1960.

Source: Special Collections Library, Pennsylvania State University. Courtesy of Mark Hajjar.

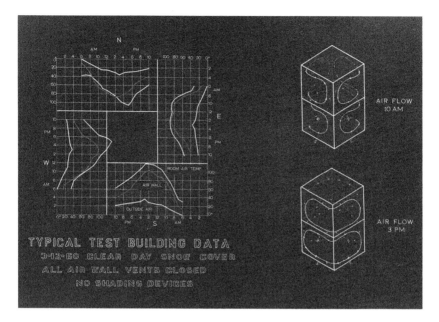

FIGURE 11.5   Measurements on March 12, 1960, showing overheating in all rooms (*left*) and air circulation in the double skin (*right*).

Source: Special Collections Library, Pennsylvania State University. Courtesy of Mark Hajjar.

the indoor temperature of 72°F was already reached in the east and south rooms around 8:30 a.m., in the west room at 10:00 a.m., and in the north room at 11:00 a.m. More severe, the indoor temperature peaked in the east room at 90°F at 3:00 p.m., in the south room at 118°F at 3:00 p.m., in the west room at 131°F at 5:00 p.m. and in the north room at 105°F at 5:00 p.m. The temperature in the DSF was lower than the indoor temperature, except in the northern DSF between 8:30 a.m. and 2:00 p.m. A diagram adjacent to the charts of the test results visualized the air circulation in the DSF chamber with air vents closed between the two floors of the air wall. The report states that a "double height air flow pattern of similar slope develops when the air wall vent between floors is open. The relative merits of having this vent closed or open between floors has as yet not been determined" (Figure 11.5 *right*).[10] It is not on record, however, whether the flow direction was observed by means of "tissue paper flags"[11] or only deduced from the temperature measurements.

These results were simultaneously promising and problematic. While they illustrated that air circulated well in the DSF, they also showed that overheating occurred even on a cold day. However, since the curtains had not been in place at the time of the measurements, no final conclusions could be drawn from the results. The missing curtains delayed further experiments until March 1961, when

FIGURE 11.6 Air-Wall Test Building, second floor with white drapes (*left*), no drapes (*middle*) and black drapes (*right*).

Source: Special Collections Library, Pennsylvania State University. Courtesy of Mark Hajjar.

the engineering team conducted measurements with white fiberglass drapes to calculate heat transfer coefficients.[12] In summer 1961, the team installed drapes of different materials and colors in the DSF and behind the single glass facade on the top with the intention to evaluate the DSF's potential of reducing indoor cooling loads (Figure 11.6). Measurements of indoor air temperature taken in September 1961 showed that a "high absorptance, low transmittance black drape" in the DSF, with all vents in the DSF fully open, was the most effective "solar radiation interceptor." With this option, heat entering the room was only 4 percent of the heat entering the single-glazed top floor room, thus confirming "that the Air-Wall concept has merit in reducing solar radiation gains for cooling conditions."[13]

No further experiments were undertaken to study the potentials of solar heating in winter by creating a temperature equilibrium in the DSF's four sides. Similarly, the above described experiments to study the reduction of cooling loads in summer were not taken further. Experiments on glass products in 1963 and later removed part of the DSF and added a shading structure in front of the facility. (It is not clear when the facility was demolished altogether.) A final publication on the Air-Wall experiments was published in *ASHRAE Transactions* in 1968, summarizing the measurements undertaken for the drapes and glass alterations. The article reiterated that the "air wall construction reduces heat gain more effectively than conventional windows." It further stated that additional "light and heat reflecting films, gray plate glass and sun screens were found to be effective in reducing solar heat gains."[14] The publication did not even mention the initial intent of studying the potentials of solar heating and temperature equilibrium around the building.

Exploring Diverse Scales

Hajjar's passion for the Air-Wall concept seemed to be driven by more than the need to contribute to industry-funded research at Penn State. Since large glass facades were used in both solar and modernist architecture, his interests in advanced technology and aesthetics were equally fueled. The virtues of transparency were most controversially discussed after Ludwig Mies van Rohe had completed the Farnsworth House in 1951, less than a decade before the Air-Wall Test Building.[15] The test facility's nickname, "the glass house," spread in non-scientific journals, such as *Popular Science* (1959) and *Popular Mechanics* (1961),[16] was certainly agreeable to Hajjar. In addition, his preference of square or double-symmetric floor plans made him favor that all facades of a building have the same aesthetic appearance.

As several of his designs show, he believed that the Air-Wall could be employed on all scales, from small residential to large office buildings. With his associate Harlan Wall, he combined the idea of an outer air-blanket zone with the idea of a service core to house kitchen, bathrooms, staircase, mechanical room, shafts for plumbing, electric, heating or cooling and other service spaces.[17] Two examples are their Heart House and Future Concept of Air-Wall Core Building (Figures 11.7 and 11.8). In the Heart House, the two principles of "core" and "air-wall" led to an entirely open plan, as the house has no interior walls touching the exterior DSF. The core includes staircase, kitchen, bathrooms, laundry,

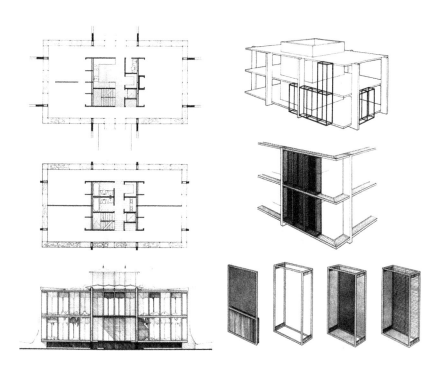

FIGURE 11.7   A. William Hajjar and Harlan J. Wall (no date), "Heart House": first and second floors, elevation and Air-Wall system.

Source: Special Collections Library, Pennsylvania State University. Courtesy of Mark Hajjar.

FIGURE 11.8   A. William Hajjar (no date), "Future Concept of Air-Wall Core Building."

Source: Special Collections Library, Pennsylvania State University. Courtesy of Mark Hajjar.

mechanical space and an elevator. Living, dining and multi-purpose room on the first floor, and four bedrooms on the second floor are organized around the core. The envelope is a four-sided double skin, consisting of floor to ceiling 2-feet deep box frames. An elevated ground floor allows air to enter at the bottom of the double skin. The Future Concept of Air-Wall Core Building is a futuristic scheme with ovoid building structures. In 2013, Hajjar's son Mark compared his father's Air-Wall idea with the atmosphere around the Earth:

> The sun heats the air on the surface of the earth that faces the sun, and the air currents circulate the air around the earth thus bringing the cool air from the unheated side of the earth to the sunny side and the heated air to the cooler side of the earth. If this happens naturally, what if a building were built with a blanket of air held in place between two walls of glass. Would the air move around the building in a similar way as it does around the earth thus heating and cooling the building without the need for any man made mechanical systems.[18]

While Hajjar's ambitions regarding solar architecture can be contextualized within the discourse of the 1940s to 1960s, there are also major differences: Almost all of the design schemes at that time emphasized the importance of south orientation and included large glass walls only on the south side.[19] With the Air-Wall, Hajjar managed to combine the idea of solar orientation with a double-symmetric floor plan, justifying his scheme by referencing the spherical, orientation-less Earth. Even if a house did not rotate like the Earth, the air would still move due to pressure differences in the double skin.

Hajjar was not able to realize any of these designs. On a much smaller scale, he included winter gardens in residential competitions, such as the Indianapolis Home Show Competition, and built a south-facing winter window in the swimming pool building of his own 1962 single-family house in State College. The discrepancy between the large attention that he stirred and the little result is astonishing. Several letters sent between Robertson Ward and Bill Hajjar show that SOM was highly interested in DSF systems at that time. In a letter from June 9, 1959, Ward confessed

> that our Inland Steel Building in Chicago was originally designed on this principle with a plane of glass at the outer surface of the 4' deep exposed columns in addition to the present glass line at the inner faces of these columns. However, the time required to properly develop and engineer the system prevented its incorporation in this building.[20]

SOM's Inland Steel Building shows the complexity of the entire endeavor of implementing DSFs at that time. Located in the densest district of Chicago, surrounded by skyscrapers, the DSFs would have been oriented only to the east and west, with little solar gain possible. The benefit would mainly have been an increased insulation of the facade. From existing facade technologies to four-sided double skins that exchange air in interconnected air chambers seemed to be a long way.

Hajjar was not the first architect to come up with a four-sided double-skin concept. Two beautiful examples are the 1903 Steiff factory in Giengen an der Brenz and the 1930 municipal natatorium in Berlin-Mitte, both in Germany. The factory's three-story eastern wing, likely designed by Richard Steiff, is well known for being the first built DSF, mainly employed for improving daylighting in the production building.[21] The main intention for employing DSFs in the natatorium, designed by Carlo Jelkmann and Heinrich Tessenow, was to increase thermal comfort and to avoid condensation on the inner glass panes.[22] Both buildings benefitted in winter from the reduced air infiltration through the facade and from solar gain, but there is no record of conceptualizing the movement of air from the sunny side of the building to the shady side.

The late 1950s and early 1960s showed an increased interest in testing the solar potential of (south-oriented) double skins. A widely discussed project was St. George's County Secondary School in Wallasey, England, completed in 1961, shortly after the Air-Wall Test Building. Designed by Emslie A. Morgan, the south facade consisted of two glass layers 24 inches apart. Reyner Banham, in his 1969 *The Architecture of the Well-Tempered Environment*, called the building's DSF a "solar wall," not only because of the two glass layers but also because it included "opaque panels, painted black on one side, polished aluminium on the other." These elements could be turned in order "to provide a degree of thermal control by absorption/reflection of solar heat." In addition, "white-painted wooden shutters" could be installed "to reduce absorption of solar heat." Banham also noted that the primary concrete structure with its high thermal mass was beneficial for storing the heat from "three main sources: the solar wall, the electric lighting, and the inhabitants."[23]

The End of the Air-Wall

The Air-Wall research team seemed to be quite uninterested in the problem of solar heat storage, a central topic for the research teams at MIT, NYU and Princeton (Lawrence Anderson, Hoyt Hottel, Aladar and Victor Olgyay, Maria Telkes, among many others).[24] Instead, Hajjar's wall sections (Figures 11.2 and 11.3) suggest that the copper-colored "radiant curtain" was intended to play a major role in the Air-Wall system. When the sun shone, it was supposed to work as a "permanent solar energy absorbing screen," which Hajjar apparently believed to be sufficient for some solar heat storage. Without the sun, the curtain became an "electrical resistive heating element" to cover the heating needs in the building "without a complex system of ducts and vents."[25] Such a super curtain product, however, did not exist. Similarly, the "possible new glass" with "reflective inner surface" (Figure 11.3), needed for the outer DSF layer to keep the electrical heat inside, was not yet developed. Relying on future products for heat storage, electrical heating and glazing turned out to be the biggest mistake of the project. The engineers on the team might have seen that more clearly, apparent in their shift to studying the efficiency of drapes in reducing cooling loads (Figure 11.6). Paradoxically, their focus on the summer case with vertical cavity ventilation did not need four sides of double skins at all.

By 1963, Hajjar backed down from his research trajectory and professorial career at Penn State and began looking for new opportunities for reasons that seemed to be related to his unfulfilled aspiration to become the head of Penn State's Department of Architecture. Other reasons could be found in the dynamics within the research team that never really followed through with investigating the air movement, the potential temperature equilibrium and the controllability of the four-sided DSF. After working as senior designer at Vincent Kling Architects and design director of Harbeson, Hough, Livingston & Larsen, both in Philadelphia, PA, he started in 1965 a new career as a developer in La Jolla, CA, where he lived until his death in 2000. Trying to revive the Air-Wall research in the late 1970s, he registered a business in Malibu, CA, under the name "Air Wall Research & Development." The problems described above returned. In 1978, he received the following advice from the consulting engineer Charles E. Duke:

> The Airwall configuration [. . .] promotes temperature equilibrium front to back and top to bottom. The equilibrium temperatures, however, were too

high in the afternoon and too low at night viz. 110°F. and 40°F. The "average" temperature 75°F. suggests that given enough thermal storage it might be possible to "heat" comfortably an Airwall building even during central PA winters. Thermal storage is usually accomplished using an "active" system which depends on controls (thermostats) and pumps (liquid systems) or blowers (air systems with either liquid or rock storage) [. . .]. Passive systems, eg.—Trombe wall, depend on mass only [. . .]. Trombe walls would reduce the 110°—40°F cycle in the Airwall structure but would not eliminate it. [. . .] There are storage media that will reduce the above problem, they are called eutectic salts.[26]

For Duke, the lack of heat storage was the major problem of Hajjar's Air-Wall. Rejecting active systems (because of the required controls, pumps and blowers) and passive systems in the form of thermal mass (because of inefficiency), Duke proposed the use of phase-change materials in the DSF. In a sketch, he presented rib-like "eutectic salt trays" on each floor of the DSF, around which air could circulate and thus "assist the transfer of heat to and from the air" (Figure 11.9). His letter is the last record for the Air-Wall project.

Shortly after that, in 1980, the completion of the Hooker Building in Niagara Falls, NY, showed that architects and engineers continued to believe in the functioning of the four-sided DSF.[27] It took another 17 years for the next attempt to build a four-sided DSF in the Götz Headquarters in Würzburg, Germany, in 1997.[28] Both buildings integrated sophisticated horizontal louvers for shading and

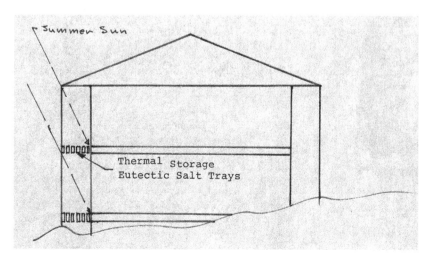

FIGURE 11.9 Charles Duke 1978, proposal to install trays with phase change material in the DSF.

Source: Special Collections Library, Pennsylvania State University. Courtesy of Mark Hajjar.

daylight redirection in the DSF. And both of them used the DSF complementary to their respective HVAC system rather than focusing on solar heat storage.

Notes

1. "The Pennsylvania State Libraries Provide an Overview of Hajjar's Work," https://libraries.psu.edu/about/collections/hajjar-heritage.
2. Interviews with James Alexander, Boston; Barry Eiswerth, Philadelphia; and Louis Inserra, Port Matilda, in 2017 and 2018.
3. This chapter is the result of a larger research project on the architect A. William Hajjar. Researchers beyond the authors include Henry Pisciotta, Laurin Goad Davis, Moses Ling, David Goldberg, Donghyun Rim, Homeira Mirhosseini and Gen Pei. We are thankful for the generous funding from the Penn State Raymond A. Bowers Program for Excellence in Design and Construction of the Built Environment and the Penn State University Libraries.
4. In "New Swimming Pool Dedicated," *The Tech* 60, no. 30 (June 4, 1940): 3, the glass facade is described under the headline "Gigantic Sun Window" as follows: "Double glazing up to a height of ten feet along the large window will enable warm air to flow between the inner glass screen and the window itself and to temper the cold air from the surface of the window." Cf. also "Sports Buildings," *Architectural Record* 89 (February 1941): 68–71.
5. Cf. Lawrence Anderson, Hoyt Hottel and Austin Whillier, "Solar Heating Design Problems," in *Solar Energy Research,* ed. Farrington Daniels (Madison: University of Wisconsin Press, 1955), 47–56. Cf. Austin Whillier, "Principles of Solar House Design," *Progressive Architecture* 36, no. 5 (May 1955): 122–26. For a longer discussion of the MIT Solar Energy Fund and its projects, see Anthony Denzer, *The Solar House. Pioneering Sustainable Design* (New York: Rizzoli, 2013); John Perlin, *Let It Shine. 6000 Years of Solar Architecture and Technology* (Novato: New World Library, 2013); Daniel Barber, *A House in the Sun. Modern Architecture and Solar Energy in the Cold War* (New York: Oxford University Press, 2016).
6. A. William Hajjar, *Proposal for a Research Grant on Air-Wall Construction* (Special Collections Library, Pennsylvania State University, March 1958), 1.
7. Ibid.
8. "Test of Air-Wall Construction," *Centre Daily Times,* August 10, 1960.
9. A. William Hajjar, *Research Proposal for Extension of Air-Wall Construction Grant* (Special Collections Library, Pennsylvania State University, June 24 and November 12, 1959).
10. A. William Hajjar, *Results of Qualitative Exploratory Tests* (Special Collections Library, Pennsylvania State University, no date).
11. Hajjar mentioned the observation of air direction with "tissue paper flags" in his proposal *Air-Wall Research Grant: Partial Summer Program* (Special Collections Library, Pennsylvania State University, June 15, 1960).
12. The results were reported in: Engineering Experiment Department, *Heat Transmission Air Wall Research Grant* (Special Collections Library, Pennsylvania State University, May 26, 1961).
13. Everett R. McLaughlin, *Thermal Performance of Air Wall Construction for Pittsburgh Plate Glass Company* (Special Collections Library, Pennsylvania State University, October 5, 1961), no page.
14. L. F. Schutrum, J. L. Stewart, R. D. Borges, and Vincent L. Pass, "Air Wall Construction—A Means of Reducing Air-Conditioning Loads," *ASHRAE Transactions* 74 (1968), part 1, I.2.1-I.2.14, here I.2.10.
15. Cf. Elizabeth Gordon, "The Threat to the Next America," *House Beautiful* (April 1953): 126–30, 250–51. Joseph A. Barry, "Report on the American Battle Between Good

and Bad Modern Houses," *House Beautiful* (May 1953): 172–73, 266–73. For discussion, see Ute Poerschke, Henry Pisciotta, David Goldberg, Moses Ling, Laurin Goad and Mahyar Hadighi, "Making the Glass House Habitable: The Debate on Transparency and A. William Hajjar's Contribution in the Mid-Twentieth Century," in *Proceedings of the Façade Tectonics 2016 World Congress* (Los Angeles, CA: Tectonic Press, 2016), 291–99.

16. "Glass Blinds Form Building Wall," *Popular Science*, June 1959, 98. "Glass 'House' Tames Solar Heat," *Popular Mechanics* 116, no. 1 (July 1961): 73. Also "Test of Air-Wall Construction." William Faust, "The Designing Professor," *The Pittsburgh Press*, March 20, 1960, 6–7. "Penn State Architect Wants to 'Bottle' Us in Glass Shell House," *Patriot News*, September 27, 1959.

17. The "core" principle might have its roots in Catharine Beecher and Harriet Beecher Stowe, *The American Woman's Home or Principles of Domestic Science* (Boston: A.H. Brown and Company, 1869). It could be found, for example, in Buckminster Fuller's Dymaxion House of 1927 or the Keck houses.

18. Mark Hajjar, *Letter to Tim Pyatt*, September 20, 2013, Special Collections Library, Pennsylvania State University.

19. The recommendation for large glass surfaces oriented to the south can be found in popular magazines, such as *House Beautiful*; industry brochures, such as the 1944 *Solar Houses, an Architectural Lift in Living* and the 1947 *Your Solar House*, both funded by Libbey-Owens-Ford Glass Company; and architectural journals, such as *Progressive Architecture*.

20. Robertson Ward, *Letter to A. William Hajjar*, June 9, 1959, Special Collections Library, Pennsylvania State University.

21. The facade, however, led to problems of glare and overheating, which was solved at some point by painting the glass.

22. Other swimming pools utilized single-sided DSFs, too, for example in Frankfurt-Fechenheim, designed by Martin Elsässer in 1929.

23. Reyner Banham, *The Architecture of the Well-Tempered Environment* (Chicago, IL: Chicago University Press, 1969), 281–82.

24. Cf. Denzer, *The Solar House. Pioneering Sustainable Design*; Perlin, *Let It Shine. 6000 Years of Solar Architecture and Technology*; Barber, *A House in the Sun. Modern Architecture and Solar Energy in the Cold War*.

25. "Test of Air-Wall Construction."

26. Charles Duke, *Letter to Bill Hajjar*, October 25, 1978, Special Collections Library, Pennsylvania State University.

27. Jim Murphy, "Rainbow's End," *Progressive Architecture* 61, no. 4 (1980): 102–5. Michael Wigginton and Jude Harris, *Intelligent Skins* (Oxford: Butterworth-Heinemann, 2002), 163–68.

28. Thomas Lödel, "Verwaltungsgebäude in Würzburg," *Detail: Solares Bauen/Solar Architecture* 3 (1999): 448.

12

DEFINING THE DOUBLE-SKIN FACADE IN THE POSTWAR ERA

Mary Ben Bonham

The Occidental Chemical Building in Niagara Falls, New York (Figure 12.1) is widely recognized as the first double-skin glass facade building to be constructed and occupied in the United States. In 1978, when the oil, gas and chemical company headquarters was commissioned, the "sealed and conditioned glass box" had become the default building type for corporate American architecture. In the wake of the 1973 and 1979 oil crises, architects and engineers were spurred to achieve higher energy performance in buildings with building enclosures as primary areas of research and experimentation. With a minimalist glass and aluminum curtain wall, Cannon's design united aesthetics and ideals of modernist architecture with the passive solar movement, boldly scaling up fundamental environmental design principles to large-scale commercial application.

With an innovative hybrid of passive design and automated systems, proper functioning of the double-skin facade relied on centralized, computer-controlled adjustment of airflow dampers and daylight-diffusing louvers positioned inside the multistory cavity. Overlooking the Niagara River Gorge, the iconic high-tech glass cube represents the first example of what an energy-efficient office building *should* look like in the eyes of postwar America. This precedent, however, eclipses a fuller history of double-skin facades that encompasses a variety of architectural expressions. A brief history of double-skin precedents through the end of the postwar era provides an expanded definition of double-skin functionality and aesthetics, including the degree to which facade systems are tectonically integrated with building heating and cooling systems. This history upsets any notion of a direct line of development of double-skin facade technologies and challenges the assumption that all double-skins must be highly complex or follow a specific aesthetic model.

FIGURE 12.1   Occidental Chemical Building (1978–81), Niagara Falls, New York. Cannon Design.

Source: © Cannon Design.

Double Skins

Double-skin facades are essentially two glazed building skins separated by an air space. Over time, in pursuit of higher energy performance and occupant comfort, double skins have adopted emerging technologies, expanding the typology into a myriad of formal solutions. Instigating reasons for constructing enclosures with double layers of glass are multifaceted and practical. When glass is desirable for light or views, the buffer zone formed by the two layers improves interior thermal comfort and acoustic isolation. The air space provides a protected location for solar shading devices. Through strategic positioning of ventilation openings in one or both layers, air flow through the double-skin cavity may be used to facilitate ventilation, solar heating, cooling or a combination of these functions.[1]

Some double-skin designs operate completely passively, using principles like the stack effect and thermal mass. Other schemes rely on intelligent automation of facade elements and varying degrees of integration with other building systems. The introduction of active systems into double-skin facades follows the advancement of active environmental controls into mid-century modern architecture. A modernist construct promoting active walls can be traced back to Le

Corbusier's *mur neutralisant* and *respiration exacte*, concepts developed in parallel with the emergence of central air conditioning. The oil shocks of the 1970s pushed research toward new forms of energy-efficient architecture into the mainstream. However, solar and double-skin architecture were by no means original to this era. A variegated progression of early and postwar modern projects built on the climate wisdom of vernacular building styles to capitalize on technological developments in building materials and systems.

Early Precedents

Traditional sun spaces like glazed verandas and balconies expand the building enclosure into occupiable, climate-adaptable zones. A variety of reconfigurable openings and coverings in the layers provide adjustable levels of light, solar heating and ventilation. Because of their adaptability, sun spaces are found in vernacular buildings in a range of cold and warm climates. Similar functionality is achieved with double window construction, in which two single-glazed windows are placed parallel in a wall opening. The air space, and extra mass, of double windows offered improved thermal comfort and better acoustic isolation than a single layer of glass. Double windows were staple elements of traditional houses in colder climates such as Germany and Switzerland.[2]

A double window system was adopted as a practical solution to enhance thermal comfort in more luxurious early modern residences throughout Europe and North America. Frank Lloyd Wright's Westcott House in Springfield, Ohio (1906–09), integrates double casement windows in long rows under deep roof overhangs.[3] Occupants could manually adjust the position of one or both hinged glass sashes for a variety of environmental effects on ventilation and temperature. Wright designed similar double frames for the Robie House windows in Chicago, Illinois (1908–10); here, hinged inner sashes held insect screen instead of glass.[4] An extra window added a layer of cost along with comfort. Single-layer windows sufficed for the majority of applications, with performance necessarily augmented by traditional means such as layers of heavy and sheer curtains, awnings and shutters.

Introduction of Active Mechanical Systems

One of Le Corbusier's earliest structures, the Villa Schwob in La Chaux-de-Fonds, Switzerland (1916–17), incorporated double windows with radiant heating pipes located in the gap between two glass layers.[5] Heating elements are concealed from view within the double-story living room window assembly. In the bedrooms, wall-mounted radiators are visible below window sills; it is probable that heating pipes extend into these generously sized glass cavities as well.[6]

Villa Schwob's windows are not so much an invention as a new application of existing technologies. It is quite common to locate a radiator under a window to take advantage of natural convection to heat the space and prevent downdrafts of cool air. Wright encased radiators within wood-paneled knee walls below ribbon windows at the Westcott and Robie Houses.[7] The influence of Frank Lloyd Wright's early houses on the young Le Corbusier has been well advanced by historians who, for example, note Wrightian qualities in Villa Schwob's spatial arrangement. Both architects employed mechanical systems alongside passive environmental principles in inventive ways that permitted their designs to depart from tradition.[8] While Wright's work surely influenced the design of Villa Schwob's fenestration, we cannot surmise the exact origin of Le Corbusier's decisions about mechanically conditioned double windows. Double windows are native to Switzerland, after all. The introduction of heating elements *inside* a glazed cavity instead of *adjacent* to it is not a far step to take. In 1914, the Expressionist writer Paul Scheerbart rightfully cautioned that heating elements within double glass walls would cause excessive thermal losses.[9] Nonetheless, the strategy persisted in Le Corbusier's oeuvre for a discrete period of time.

Le Corbusier went on to envisage mechanically conditioned wall cavities for buildings of significantly larger scale and complexity of program. For the Centrosoyuz office building in Moscow, a competition project awarded to Le Corbusier and Pierre Jeanneret with Soviet architect Nikolai Kolli in 1928, he proposed a double enclosure for the entire structure: walls, roof and underside of the first floor supported by *pilotis*. The system he called *mur neutralisant* would circulate warm or cool conditioned air (according to the season) into sealed wall cavities, in combination with *respiration exacte*, a centralized mechanical system to supply purified and tempered ventilation air to building interiors at a constant 18° C. Seeking validation for the unusual scheme, Le Corbusier had the design evaluated by the American Blower Company, who concluded that the concept would consume an undue amount of energy. In the end, conventional radiators were installed at Centrosoyuz.[10] Stone walls were built as single wythe, without the benefit of even a passively insulating air cavity. Glazed facade areas *were* constructed in double layers, with unconditioned cavities and rows of operable sliding windows—precedent for ribbon double windows existing in the recently completed Narkomfin housing block in Moscow (1928–32) by Moisei Ginzburg and Ignatii Milinis. On primary facades, glass curtain walls stretched over the height of six- and seven-story office blocks. Translucent glass was used in upper lites in the inner skin of these facade areas for added solar control; however, thermal comfort in Centrosoyuz was never effectively realized.[11]

Concurrent with Centrosoyuz, Le Corbusier was working on another large-scale commission, the Salvation Army Building in Paris, France (1929–33). This design featured a south-facing glass curtain wall that enclosed five floors of residential quarters and a nursery. The *mur neutralisant* and *respiration exacte* combo proposed for this project was rejected due to uncertainty and cost. Testing over a period of two years by French glass manufacturer Saint Gobain confirmed the earlier reply; the *mur neutralisant* would require excessive energy to operate effectively. Saint Gobain engineers finally reported that the system could work if a third layer of glass was added, with still air trapped between the outer two layers—e.g., a double-pane unit in the outer window. Certainly, insulated glazing unit (IGU) technology, under development at the time in the United States, would have been an improvement to the concept. The facade was built as a sealed and unshaded single-layer glass wall behind which conditions proved unbearably stifling. After the first summer, Le Corbusier was ordered to have sliding windows installed in the upper third of each section of the curtain wall. Le Corbusier's concept for mechanically active double walls was devised not so much to conserve energy as to improve thermal comfort. The problem of thermal losses through glazing could be, in his view, resolved through integration of newly available centralized air-conditioning technology. The curtain wall was damaged by bombing during World War II, after which Le Corbusier had the opportunity to redesign the facade. This time the architect turned to more architectural, less mechanically reliant *brise-soleil* as a way of conditioning the window wall.[12]

Mid-century Developments

At mid-century, glass had assumed connotations beyond the basics of admitting light to see. Glass architecture expressed modernity and a healthful connection to the natural world, and the onset of World War II ushered in material and fuel rations. In this context, glass manufacturer Libbey Owens Ford developed Thermopane, beginning factory production of the product in 1937. Double glazing became a useful strategy to open homes and workplaces more fully and comfortably to daylight and views by reducing thermal losses, downdrafts and condensation. The work of Chicago architect George Fred Keck is a prime example of developments in glass enclosures that began before World War II and gained traction during and after the war. In the 1940s and 1950s, Keck's designs for "solar houses" featured large, sun-facing Thermopane windows.[13] Manufacturers worldwide began offering a steady stream of products with improved performance through better spacers and seals, and new glazing such as float glass and heat-absorbing glass. Window frame technologies likewise advanced. New materials and profiles with thermal breaks enabled larger window openings.

In the United States, interest in solar homes declined in postwar years. Concerns over a secure energy supply receded, and expansive glass walls and IGUs were considered costly luxuries in suburban home development.[14] Double glazing gradually came into more common use in Europe, especially where double windows would have been an option. Two buildings by Josse Franssen in Brussels, Belgium, are a case in point. In 1950, a Franssen-designed apartment building was built with large double windows on south and west facades; sliding adjustable louver panels were located between the two window layers. A few years later, Franssen shifted to specifying Thermopane for a similar apartment building.[15]

Double-Skin Glass for Industry

Whereas large volumes of glass were a luxury for residential settings, the scaling-up of windows to full walls was an imperative project for the modernization of factories. The aim was to admit more light, deeper into workplace interiors. The dissolution of the massive wall had already occurred by the beginning of the twentieth century thanks to developments in cast plate glass and cast-iron framing. The Steiff toy factory building (Figure 12.2) built in 1903 in Giengen, Germany, was a pioneering but isolated incidence of fully glazed double-skin facade architecture. The factory's three-story glass and iron enclosure was conceived to provide thermal comfort while admitting plentiful daylight through large expanses of glass. A single-glazed outer skin, stretched as a continuous plane across the full height of the iron structure, is known as the first glazed curtain wall in Europe. The outer skin is fixed about 25 centimeters in front of a floor-to-ceiling height single-glazed inner skin. Matte finish on the glass and interior draperies served to reduce heat and glare. While the air space was not ventilated, box-type windows that punched through both layers of the facade permitted cross-ventilation of the factory floor where the popular Steiff sitting teddy bears were made.

The Steiff factory's innovative enclosure did not directly influence architects at the time—the building was not in a major city nor designed by an architect, leaving it little chance to be noticed. Furthermore, the glass curtain wall had not yet become an acceptable form of enclosure.[16] This was soon to change. Expansive areas of glass facade were applied with similar inventiveness in single-skin form to

FIGURE 12.2 Steiff Factory (1903), Giengen, Germany. Eisenwerk München AG.

Source: Zacharias L. (cc-by-sa/3.0).

frames of steel and reinforced concrete at the AEG Turbine Factory by Behrens (1908–9) and the Fagus Factory by Gropius & Meyer (1911–13).[17]

Fully glazed enclosures were still considered avant-garde when a double-skin was applied in 1939 to an Italian factory building. Adriano Olivetti commissioned Luigi Figini and Gino Pollini to design a series of enlargements to Olivetti's workshop complex in Ivrea. The architects embraced the spirit and techniques of international avant-garde architecture. Factories were laid out according to the needs of newly modernized production lines and structured with reinforced concrete frames that effectively opened up facades to extensive fenestration.

In the first enlargement (1933–36), ceramic tile-clad single-skin facades feature continuous ribbon windows. The second enlargement project (1937–39) consisted of rear additions and planning for a third, more significant enlargement. The third enlargement (1939–40) was a 130-meter-long 3.5-story structure. The latter building's north-facing facade, fronting Via Jervis, was enclosed by two parallel glass planes. Iron-framed glazing in both inner and outer skins is distinguished by a regular pattern of 3-meter square window modules with a combination of fixed and operable sashes. Arrays of pivoting hardboard panels are positioned inside the facade's 50-centimeter-deep cavity for comprehensive solar control. South-facing walls of the factory took a single-skin approach. Glazed walls here were fitted with sections of concrete *brise-soleil* and adjustable horizontal metal louvers.

The Italian architects were fully aware of Le Corbusier's concept of conditioning an air cavity between two layers of glass. They rejected the active wall concept in favor of a passive approach.[18] The genius of the Olivetti double facade lies in manually operable sashes which allow the space to be closed when heating is desired or opened to evacuate excess heat passively through the buoyancy-induced movement of solar-heated air. Whereas most other modernist glass facades had controlled heat and light using venetian blinds, draperies or other types of screens on the building interior, Figini and Pollini saw the wisdom of placing operable solar control *between* two glass skins before heat can enter the conditioned space.

A fourth enlargement to the Olivetti workshop complex built in 1956–57 repeats the double-skin glass wall format, creating an even more extensive presence of gridded glass along the Via Jervis (Figure 12.3). The extent of this latest addition can be identified by the introduction of concrete-framed planter boxes and vertical piers of yellow and white ceramic tile.[19]

The Olivetti facade first realized in 1939 represents an early and successfully enduring instance of ventilated double-layer glass facade with solar control in the cavity. The use of operable shading devices encased in the Occidental Chemical office building's double-skin facade makes it a descendant of the Olivetti prototype, with the prime distinction that ventilation through the Occidental's facade cavity was purely external; there were no ventilation openings in the air-conditioned building's inner skin. With its provisions for natural ventilation of interior workspaces, the Olivetti is a more direct precedent for the ventilated

FIGURE 12.3   Olivetti Workshops, Ivrea, Italy. Figini and Pollini. 1939–40 addition in background; 1956–57 addition in foreground.

Source: Laurom (cc–by–sa/3.0).

double skins developed in the 1990s for "ecological office buildings" like the Commerzbank and RWE towers in Germany.

Post-World War II Developments

After the war, glass curtain walls were increasingly applied to fully air-conditioned buildings. As office buildings in this period became more internally load-dominated, thermal heating was less of a concern and solar control could be handled by increasing the capacity of mechanical systems. Despite this, memories of war-time rationing were fresh, and significant research continued related to conservation of energy and materials. Representative of a regional approach, Victor Olgyay's philosophy of bioclimatic architecture emphasized passive principles of solar orientation, external shading, natural ventilation and thermal mass.[20]

Even as the passive solar movement remained on the margins of architectural production, technologies foundational to double-skin facades were developed in the postwar decades that preceded the energy crises of the 1970s. As was the case for the first half of the century, developments pushed both passive and active technologies into new directions. Some projects attempted more porous designs that permitted natural ventilation of the interior or hybrid solutions that combined passive, active and automated strategies. Other projects effectively realized fully sealed enclosures integrated with building interior air-conditioning systems. Double-layer construction was not common; however, a number of isolated inventions made lasting contributions that would impact future designs.

Passive Double Enclosures

In 1961, a ventilated double-skin glass façade was applied to the south-facing wall of St. George's School in Wallasey, United Kingdom. The design by architect Emslie Morgan employed a combination of passive techniques, including thermal mass, moveable insulation and shading devices and cross-ventilation.[21] As in the Olivetti facade, kinetic components of the wall required a daily routine of manual operation.

A different type of naturally ventilated double wall technology was developed in France by engineer Félix Trombe and architect Jacques Michel. In this concept, a glass wall is positioned in front of an opaque thermal mass wall. Prototypes of this "indirect gain" solar collector were built in 1967 and 1974.[22] Trombe walls were less commonly employed in residential design than indirect gain spaces based on the concept of a greenhouse. South-facing glazed sun spaces were adopted into the vocabulary of grassroots solar architecture in the United States and Europe. In some experimental houses, sun spaces were paired with other passive strategies such as "double envelope" construction, earth tubes, super insulation and thermal storage in rock, dirt or water.

Double envelope homes built in the 1970s relied on passive solar energy and convective heat flow to drive circulation of air. In a typical design, a continuous insulated gap is framed for air to flow in a loop connecting a south-facing "solar engine" greenhouse space, an air gap following the roof line, an air gap across the full height and width of the north wall and a thermal storage area below the house with an air connection back to the greenhouse. Windows in this home type, which enjoyed a niche level of popularity in United States the 1970s and 1980s, were constructed with glass double windows on opposing sides of the typically 12-inch deep air space.[23] This contrasts with one of the first known double envelope structures, the Loomis House by William Lescaze in New York State (1937), in which a separate air-conditioning system for the air space was designed to insulate and prevent condensation on glass surfaces even while the interior was maintained at a high humidity for the health of its occupants.[24]

Active Double Enclosures

Other projects during the postwar period incorporated active systems alongside passive principles. The office of Ove Arup, an engineer, worked with the architectural office of Michael Scott to design the Store Street Bus Station in Dublin,

Ireland (1945–53). Called "Busáras" by Dubliners, the city's main bus terminal with offices above (Figure 12.4) is a functioning realization of Le Corbusier's *mur neutralisant* concept. On office elevations, the outer skin is fully glazed and the inner skin consists of bronzed-framed windows above a knee wall. The 18-inch-deep facade cavity is fitted with radiant heating pipes. In addition to improving thermal comfort adjacent to the large windows, the design aimed to provide occupants with an acoustic buffer from the traffic center below.[25]

An architect and professor at Penn State University, William Hajjar, developed a glazed double-envelope concept that employed a "radiant curtain," an active polyvalent control layer within the cavity. With the support of a grant from Pittsburgh Plate Glass Company, Hajjar built a test building with a two-story "Air-Wall" (1959–61).[26] Like the Olivetti facade, the Air-Wall had operable vents to exhaust heated air from the cavity, but unlike the Olivetti facade, the Air-Wall did not permit ventilation through the facade to the interior.

Developed in northern Europe during the 1960s, airflow windows represented another category of double windows. A typical airflow window consists of a double-glazed outer sash and a single-glazed inner sash with adjustable blinds in the cavity. The two sashes share a thermally broken frame. An airflow window can be either passive or active depending on its configuration. In the passive "exhaust-air" type, conditioned room air is circulated upwards or downward through the window cavity to be released through the outer sash. In the active "return-air" version, room air passing through the window cavity is returned via ductwork for energy recovery in the building's central air handling system. An airflow window patent was filed in Sweden in 1956. Airflow windows designed by the engineering company Ekono were first applied in a large-scale project

FIGURE 12.4   Busáras (1945–53), Dublin, Ireland. Michael Scott with Ove Arup & Partners.

Source: Jnestorius (public domain).

in the City of Helsinki's Building Department offices in 1967. By 1982, airflow windows had been applied to over 50 large buildings in Europe.[27]

The concept of an active return-air airflow window was scaled up to a full window wall for the British Sugar Corporation building in Peterborough, England (1972–73). Designed by Arup Associates, the office building incorporates a two-story double-skin glass facade. A sealed exterior skin was an essential component for this location to keep out noise and dirt stirred up by the adjacent beet sugar processing operations. Conditioned interior air is returned through the facade cavity to buffer against external temperatures and sounds.[28]

Introduction of Automation

The postwar passive and active innovations in double-skin enclosures—seasonal adaptability, solar control layer and adjustable airflow inside the wall cavity, energy storage and recovery via thermal mass or integration with mechanical system—formed the basis of design for the Occidental Chemical Building and other energy-efficient office buildings of the late 1970s and early 1980s. Primary signifiers of buildings at the end the postwar era as compared to earlier precedents were an unwavering reliance on air conditioning and sealed facades, along with increased use of sensors and automation to control moving parts.

In the Olivetti workshops, employees had the time-consuming task of manually adjusting windows and shading panels to achieve the desired lighting or ventilation conditions on the factory floor. In the postwar period, technologies for building automation evolved rapidly. Centralized pneumatic control was developed in the 1950s, followed by electromechanical control systems in the 1960s. After the 1970s oil crises, minicomputers and programmable logic controllers were employed to increase energy efficiency of operations.[29]

The Occidental Chemical Building

A new headquarters for Occidental Chemical materialized when the company, which purchased Hooker Chemical in 1968, intended to consolidate operations for Hooker in downtown Niagara Falls. As the building was being designed by Cannon Design in 1978, it became known that land Hooker had sold to the city of Niagara Falls in the 1940s and 1950s was severely contaminated with industrial waste. A whole community of housing and schools that had been built in Love Canal was evacuated in 1978 after the problem of toxic waste became a visibly unavoidable problem. With this inauspicious disaster in the near background, it was important for Occidental Chemical to try to polish its image. The architects envisioned an all-glass enclosure to maximize views and daylighting. Newly instituted limits on window area, a response to the energy shortages, would have limited these views. The double-skin strategy was seen as a solution to counter heat losses and gains in an all-glass building. The idea of an energy-efficient building of revolutionary design became an attractive way to signal the company's leadership in finding solutions to the global energy crisis.[30]

Such a building had not been built in the United States before. Cannon augmented its in-house team of architects and engineers, led by principal-in-charge Mark Mendell, with consultants from across the country to consult on energy and facade systems: Burt Hill Kosar Rittelman Associates of Cambridge, Massachusetts; John Yellott of Arizona State University; and Richard S. Levine of the University of Kentucky. Three alternative schemes were developed and analyzed. A "dynamic skin concept" promised the highest energy efficiency and was selected by the client. The team calculated that the building would consume 39 kBTU/ft²/yr, well below the ambitiously targeted energy consumption of less than 55 kBTU/ft²/yr. A typical energy-efficient office building in a similar climate at the time consumed around 70 kBTU/ft²/yr.[31]

A double-skin facade was central to the solution. As was the norm for office buildings at the time, the building was air conditioned; both skins were tightly sealed (Figure 12.5). The enclosure was constructed with a blue-green tinted double-glazed outer skin and a clear single-glazed inner skin. A 4-foot-deep cavity stretched the entire height and enclosed all four sides of the cube-like building (Figure 12.6). Horizontal solar control louvers mounted inside the cavity close to the inner skin were controlled by photoelectric sensors, one per elevation. The adjustable louvers (Figure 12.7), a now-standard option for energy-efficient enclosures, were not available commercially at the time. Cannon appropriated a suitable louver originally designed for controlling air in large ducts. When painted white, the aerodynamic shape proved effective at diffusing light into the office space while maintaining views. Airflow within the cavity was controlled by dampers at the base of the façade and air exhaust louvers at roof level. Time- and temperature-activated sensors guided the cavity closing or opening, enabling the facade to shift between buffer and externally ventilated modes. Facade airflow was isolated from the building's internal ventilation systems. Heat that was recovered year-round from electrically driven centrifugal chillers and building exhaust air extract was used for preconditioning ventilation air. Facade and interior building systems were united under control of the building automation system. Sensors in the facade cavity and building guided integrated, intelligent control of solar shielding, lighting, HVAC, fire alarm and security systems.[32]

Occidental's Impact

The Occidental was hailed as a model of urban renewal and energy efficiency upon its completion in 1981. Corporate architecture of the time had embraced the glass curtain wall as the enclosure of choice, relying on tinted glazing, fluorescent lighting and plenty of air conditioning to make the interiors habitable. In response to the energy crises, the larger architectural community was waking up to energy-conscious design, heretofore marginalized as a fringe movement.

FIGURE 12.5 Occidental Chemical Building. Energy-efficient controllable facade elements wrapped in a serenely regular green glass curtain wall.

Source: © Cannon Design.

Writing for *Progressive Architecture* in 1983, John Morris Dixon concludes that, "for now, Hooker remains the landmark application of the double-envelope concept—and serene proof that high energy performance need not call for unconventional forms."[33] One can almost hear a sigh of relief that this environmentally innovative structure was not visibly "counterculture." Cannon's design united the aesthetics and ideals of modernist architecture with the passive solar movement, at least for a time.

More glass, and more layers of it, was an attractive solution to maintaining the modernist vision (think Mies van der Rohe's glass tower) in a time of energy

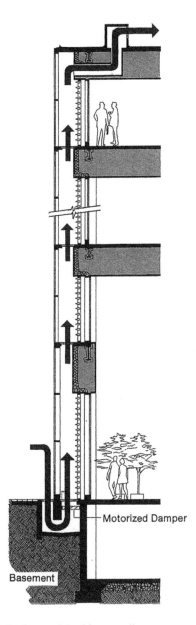

Motorized Damper

Basement

FIGURE 12.6 Occidental Chemical Building. Wall section.

Source: © Cannon Design.

FIGURE 12.7 Occidental Chemical Building. Dampers at the base and top of the multistory cavity facilitated airflow through grilled openings at each floor level; louvers in the cavity and perimeter electric lighting systems in the office spaces responded automatically to ambient daylight levels.

Source: © Cannon Design.

insecurity. The problem of poor thermal performance has been addressed by the glass industry in the form of double glazing, addition of special gasses and films, and now triple glazing. The architectural solution of a second skin offered a different kind of resolution, a tectonic problem that could be creatively addressed by collaboration between architects and engineers.

The emerging concept of intelligently responsive architecture was the hook that made solar design alluring to the mainstream. The potential to automatically modulate the air space and associated facade components has allowed double skins to dynamically adapt to a wide array of climate and use conditions. Scale is also a factor in this trend towards automation. In some buildings, window openings and coverings can be effectively controlled manually by occupants or a facility manager. In large-scale buildings, however, centralized systems provide more control over these actions and, thus, greater energy savings.

The Occidental Chemical Building was significant for the way its state-of-the-art computerized building automation system integrated control of facade elements with building mechanical and electrical systems. Automation offered tantalizingly more precise control of timing and positioning of the controlled elements. Automation and integrated control were also critical for making the concept work. Over time, the facade ceased to function in its original form. The building's decline commenced when its founding owner–occupier moved out. Its facade had been designed for centralized control; operations literally fell apart under changing management and the unplanned-for multitenant situation. The louvers stopped working in the mid-1990s and were later removed. Facade air intake grilles were covered with plywood, and seals in many of the insulated glazing units failed. With the formerly protective functions of the facade in disarray, some tenants even installed independent air conditioners inside the cavity in a quest for thermal comfort.[34] Other innovative buildings of this era had comparable building operation challenges but avoided failure thanks to stable occupation and maintenance by the same owner that had commissioned the building. Drawing lessons from the events that negatively impacted the Occidental Chemical building, Charles Linn writes, "any building has to make long-term economic sense, and exotic ones that demand special effort to work as designed need owners who are committed to maintaining them."[35]

Other double-skin facade office buildings completed soon after synthesized the facade cavity more closely with mechanical systems. In Arup Associates' Briarcliff House in Farnborough, UK (1978–83), head office for Leslie and Godwin insurance company, supply air ductwork is located within a multistory double-skin cavity. Air drawn through the facade's open base rises in the cavity via the stack effect to a mechanical penthouse level, generating a source of passively preconditioned ventilation air for the building in the winter (Figure 12.8). Another building, Lloyd's of London in London, UK (1978–86), by Richard Rogers Partnership, features an internally ventilated double skin that is a scaled-up rendition of a return-air airflow window.[36]

FIGURE 12.8   Briarcliff House (1978–83), Farnborough, UK. Arup Associates.

Source: © Given Up (cc-by-sa/2.0).

Briarcliff House uses the glazed facade as a space of energy production, but the heat engine is passive solar rather than fitted with fuel-driven heating pipes. The Lloyd's facade operates on the principle of energy recovery to reduce thermal gains and losses through glazed facade areas. The Briarcliff and Lloyd's facades are more passive versions of Le Corbusier's *mur neutralisant* and *respiration exacte* concepts, in effect integrating the two concepts within hybrid natural-mechanical systems. Facades are repositioned as mechanical plenums; conventions of floor and ceiling plenums are likewise reconsidered through technologies like under-floor air distribution and air-extract lighting fixtures. Higher levels of integration with mechanical systems was intended to amplify comfort and energy perfor-mance, with increased engineering complexity as the price tag.

Conclusion

The desire for ultimate responsiveness, for polyvalence, is captured in Mike Davies' 1981 essay, "A Wall for All Seasons." Davies writes: "What is needed is an environmental diode, a progressive thermal and spectral switching device, a dynamic interactive multi-capacity processer acting as the building skin."[37] In one sense, this objective can be criticized as stretching material limits for performance at any cost. On the other hand, this kind of visionary target setting is a driver of innovation. Over time, glazing technology has indeed evolved to extreme lengths to deliver performance, up to the present point where limits have been reached in the quest to optimize light transmittance with low emissivity and thermal conductivity. With glazing performance under static conditions reaching a peak, attention has turned to dynamic performance. Davies and Hajjar would approve of materials like daylight-responsive electrochromic glass and thermally responsive aerogels. Given the pressing need for resiliency and carbon reduction, facade solutions are needed that are effective and affordable for widespread implementation to new and existing buildings alike. The examples presented here demonstrate that the technology and the tectonics of double-skin facades remains a field ripe for continued interpretation and definition.

Notes

1. Mary Ben Bonham, *Bioclimatic Double-Skin Facades* (New York: Routledge, 2019), 15–75.
2. Eberhard Oesterle, Rolf-Dieter Lieb, Martin Lutz and Winfried Heusler, *Double-Skin Facades: Integrated Planning* (Munich: Prestel Verlag, 2001), 178–98.
3. Westcott House Foundation, "Typical Section of Wall," Archived drawing no. 0712.035, n.d.
4. "Frederick C. Robie House, 5757 Woodlawn Avenue, Chicago, Cook County, IL," Historic American Buildings Survey, documentation compiled after 1933. From Prints & Photographs Division, Library of Congress (HABS ILL,16-CHIG,33-; www.loc.gov/pictures/item/il0039/).
5. Ignacio Fernández Solla, "Le Corbusier: A French lesson on 'Murs neutralisants'," *Facades Confidential,* April 19, 2012, http://facadesconfidential.blogspot.com.
6. Marie-Eudes Lauriot-Prévost, "Univers. Villa Turque à La Chaux-de-Fonds," *Point de Vue,* January 22, 2018, www.pointdevue.fr/art-de-vivre/universvilla-turque-la-chaux-de-fonds_4588.html.
7. The radiator boxes are detailed on the *Westcott House construction drawings and the Robie House survey drawings.*
8. Reyner Banham, *Theory and Design in the First Machine Age* (1960) quoted in Paul Venable Turner, "Frank Lloyd Wright and the Young Le Corbusier," *Journal of the Society of Architectural Historians* 42, no. 4 (December 1983): 350–59.
9. Dennis Sharp, ed., *Paul Scheerbart: Glass Architecture and Bruno Taut: Alpine Architecture* (New York: Praeger, 1972), 42.
10. Rosa Urbano Gutiérrez, "Pierre, revoir tout le système fenêtres: Le Corbusier and the development of glazing and air-conditioning technology with the Mur Neutralisant (1928-1933)," *Construction History* 27 (2012), 107–28.
11. Solla, "Le Corbusier."
12. Brian Brace Taylor, *Le Corbusier, the City of Refuge, Paris 1929–33* (Chicago, IL and London: The University of Chicago Press, 1987), 111–13.
13. Daniel A. Barber, "Tomorrow's House: Solar Housing in 1940s America," *Technology and Culture* 55, no. 1 (January 2014): 1–39.
14. Ibid.
15. Stephanie Van de Voorde, "Glass and Glazing in Post-war Belgium (1945–70). The Rise of Double Glazing," in *Proceedings of the Third Annual Construction History Society Conference* (Queen's College: University of Cambridge, April 2016), 429–37.
16. Ignacio Fernández Solla, "The Steiff Factory and the Birth of Curtain Walling," *Facades Confidential,* November 27, 2011, http://facadesconfidential.blogspot.com.
17. David Yeomans, "The Prehistory of the Curtainwall," *Construction History* 14 (1998): 59–82.

18. Olivetti Historical Archive Association, "Factory Architecture in Italy—The Olivetti Workshops in Ivrea: 1896–1958," www.storiaolivetti.it/articolo/44-le-officine-olivetti-a-ivrea-1896-1958/.

19. Ibid.

20. William W. Braham, "Erasing the Face: Solar Control and Shading in Post-Colonial Architecture," *Interstices* 5 (2000): 104–13.

21. Neil S. Sturrock, ed., "St George's School, Wallasey," in *Merseyside & North Wales Region of CIBSE 75th Anniversary* (CIBSE, 2009), www.hevac-heritage.org/electronic_books/M&NW_anniversary/M&NW_anniversary.htm.

22. Colin Porteous with Kerr MacGregor, *Solar Architecture in Cool Climates* (London: Earthscan, 2005), 88–89.

23. "The Double-Envelope House," *Mother Earth News*, March/April 1982.

24. William W. Braham, "Active Glass Walls: A Typological and Historical Account," *AIA Convention*, Las Vegas, 2005, http://works.bepress.com/william_braham/2/.

25. Sarah A. Lappin and Una Walker, "Bus Transportation—Córas Iompair Éireann and Michael Scott," in *Infrastructure and the Architectures of Modernity in Ireland 1916–2016*, eds. Gary A. Boyd and John McLaughlin (Surrey, England: Ashgate Publishing, 2015), 65–86.

26. Ute Poerschke, Henry Pisciotta, David Goldberg, Moses Ling, Laurin Goad, and Mahyar Hadighi, "Making the Glass House Habitable: The Debate on Transparency and A. William Hajjar's Construction in the Mid-Twentieth Century," in *Proceedings of the Façade Tectonics 2016 World Congress* (Los Angeles, CA: Tectonic Press, 2016), 291–99.

27. Kurt Brandle and Robert F. Boehm, "Airflow Windows: Performance and Applications," in *Proceedings of the ASHRAE/DOE Conference* (Clearwater Beach, FL: ASHRAE, 1982), 361–79.

28. Historic England, "Former British Sugar Headquarters and Offices," https://historicengland.org.uk/listing/the-list/list-entry/1462710 (archived).

29. Albert T. P. So, "Building Control and Automation Systems," in *Perspectives in Control Engineering Technologies, Applications, and New Directions,* ed. Tariq Samad (Hoboken, NJ: IEEE Press, 2001), 393–95.

30. Charles D. Linn, "Ancestors of the Kinetic Facade," in *Kinetic Architecture: Designs for Active Envelopes*, eds. Russell Fortmeyer and Charles D. Linn (Mulgrave, Australia: Images Publishing Group, 2014), 12–27.

31. Cannon Design, "Hooker Chemical Company," PDF file, n.d.

32. Michael Wigginton and Jude Harris, *Intelligent Skins* (Oxford: Butterworth-Heinemann, 2002), 163–68.

33. John Morris Dixon, "Glass Under Glass," *Progressive Architecture* 83, no. 4 (1983): 82–85.

34. Terri Meyer Boake, "The Tectonics of the Double Skin: North American Case Studies," www.tboake.com.

35. Linn, "Ancestors of the Kinetic Facade," 27.

36. Leonard Bachman, *Integrated Buildings: The Systems Basis of Architecture* (Hoboken, NJ: Wiley, 2003), 166–77, 372–82.

37. Mike Davies, "A Wall for All Seasons," *Royal Institute of British Architects Journal* 88, no. 2 (February 1981): 55–57.

13

ENCLOSURE AS ECOLOGICAL APPARATUS

Biosphere 2's "Human Experiment"

Meredith Sattler

In 1948, James Marston Fitch proposed a novel conceptualization of architectural history and design—one deeply informed by materialist approaches.[1] His book, *American Building: The Forces That Shape It*, was influenced by his World War II experiences in housing and meteorology, which led him to posit that American building was undergoing a radical shift where "the genuine integration of building with science and technology"[2] resulted in a re-conceptualization of the function of architecture. In particular, he identified the building envelope as a regulator of physiological transactions between the human body and the environment. His functionalist view rendered the architectural envelope as a mediating filter that took "the load of the natural environment off man's body and thus free[d] his energies for social productivity."[3]

Simultaneously, outside the discipline of architecture, the techno-sciences that propelled World War II produced new kinds of environments, built forms, modes of production and inhabitation. With the increasing exactitude of quantification, the ability to manipulate cybernetic feedbacks and the advent of large-scale controlled experimentation within environmentally isolated laboratories, advanced scientific knowledge co-produced technologies that were entangled with social and political agendas in novel ways. Through innovations in the technocratic mechanization of warfare, such as the calibration of the cockpit to "the human body and psyche, and vice versa,"[4] the biopolitical machine of the Nazi regime and the instantaneous environmental annihilation of Nagasaki and Hiroshima, new conceptualizations and physical manifestations of containment emerged. Post-World War II, the Cold War era produced environmentally controlled life-supporting capsules that were launched into the alien environment of outer space, and existed below the backyards of American suburbia in the form of nuclear fallout shelters. The concept of *Cabin Ecology*[5] was born, and with it, the need for ever more tightly sealed envelopes.

Advances in environmental thinking have also implicated the conceptualization of relationships between envelopes and the *natural* environments they contain. In 1962, Rachel Carson's *Silent Spring* stunned the developed world by revealing the inescapability of toxic bioaccumulation, mobilized through extensive ecosystems via cybernetic feedbacks, and quantitatively expressed through the emergent sciences of Ecological Systems Theory (and later the *Gaia Hypothesis*). A decade later, we saw the entire Earth for the first time, enclosed by the blackness of space, in the Apollo 17 *Blue Marble* photograph. In it, Earth appeared complete and breathtakingly beautiful, but simultaneously lonely and isolated within, and contained by, the distinct void of outer space. Earth's tight envelope was revealed, and along with it the gravity well-induced container within which the feedback loops of natural processes cycle matter through our Biosphere. Since World War II, the properties and behaviors of these cycles have been increasingly understood through ecological and other systems sciences.

These scientific knowledges produced boundaries, technologies and politics that have conditioned contemporary understandings of environment. They facilitated our current conceptualization of the Anthropocene's *Great Acceleration*, fueled largely by the positive feedback loop of atmospheric greenhouse gases, trapped within Earth's tight atmospheric envelope. More comprehensively than any other environmental experiment conducted prior, the Biosphere 2 (B2) enclosure experiments were an ultimate expression, and test, of this new conceptualization of a materially constrained world, and its synthetic ecologies, ecotechnologies and novel life-ways.

Biosphere 2 Enclosure Missions, 1991–1994

From 1991 to 1994, B2 operated as the tightest envelope ever constructed: 1,000 times tighter than commercial building industry design specification standards of the period[6] and over 360 times tighter than the Space Shuttle.[7] Located north of Tucson, Arizona, B2 was designed to perform a "Human Experiment"[8] to test the premise that eight people could live sealed inside a materially closed but energetically open synthetic environment for two-year intervals, producing their own food and oxygen and detoxifying their wastes. B2 was conceptualized as a *mini-Earth*, 30 trillion times smaller than Earth (affectionately referred to as Biosphere 1, or B1, by project members). Their approach was holistic and innovative, synthesizing five artificial subtropical ecosystem services provisioning *wilderness* earth biomes with *domesticated* spaces for human habitation. B2's 2.5-acre photosynthetic datum was composed of six *biomes*, or hybrid ecotechnical constructions, that produced oxygen, cycled nutrients and purified air and water: a rainforest, savannah, desert, marsh, ocean and an intensive agricultural area, which collectively contained biomass from five continents. The Biospherians lived in an attached *mini-metropolis* consisting of apartments, a kitchen and common areas, including a library, gym, medical facility, communications center and analytical laboratories. Mission 1 commenced at 8:00 am on September 26, 1991, and ended at 8:20 am September 26, 1993, as the eight Biospherians walked out of the airlock, thinner and with altered blood chemistries.

Like its not so distant predecessor, the Cold War era outer-space capsule, B2 was an almost entirely isolated and materially self-contained cybernetic environment, specifically tuned to the objectives of its mission and the biological needs of its occupants. But unlike the short-term, externally provisioned, waste-storage/removal, material strategies of the space capsule, B2 was designed and operated as an entirely materially isolated system, a much more rigorous and significantly longer-duration version of *Cabin Ecology*. The Biospherians launched their missions with all the molecules they would have access to for their entire stay. With only those molecules, they were tasked with producing oxygen through photosynthesis, growing their food and recycling their wastes without polluting their environment. Through time, as their bodies, and the bodies of all the plant and

FIGURE 13.1   Biosphere 2 in 2001. Note the South Lung dome in the *upper right* corner.

Source: Photo by Lynn Douglas Lawver.

animal species with which they co-habitated, transformed their environmental biogeochemistry, their entire physio-environmental ecology began to deviate from Earth's.

Architectural Envelope as System Boundary

A feasible way to design, measure and ultimately maintain life-support systems within this shifting environmental circumstance was to conceptualize and materialize B2's envelope as a system boundary, that kept as much separation as possible between B1 and B2. The design challenges required engineering for what Bruno Latour has referred to as "The Parliament of Things,"[9] human and non-human assemblages that perform together as entangled and hybrid networks. This chapter unpacks Biosphere 2's "Human Experiment" through the lens of its enclosure, charting its innovative envelope and simultaneously revealing implications of tightly enclosed environments. B2's exceedingly tight envelope revealed novel complications and required the innovation of previously untested approaches and technologies. These kinds of implications are increasingly commonplace today: in the production of energy efficient buildings, in the recent resurgence of initiatives to colonize other celestial bodies within our solar system, and through our evolving understandings of how to live gracefully within our planet's boundaries in the age of the Anthropocene.

Designing and Constructing Biosphere 2's Envelope

Early in their design process, the Biospherians' identified three entangled reasons as to why a closed ecological system was so critical to their experiment. First, it facilitated a scientific *basic research* approach, second it allowed the use of variables to test general biospheric understandings and, finally, it generated transferrable knowledge that could be applied to the development of portable and potentially off-Earth biospheres without perpetual resupply from Earth.[10] To accomplish their research agenda, they developed an exceedingly tight envelope that functioned as a closed ecological system boundary, a physical manifestation of the systems boundaries that appeared in the numerous ecological systems diagrams they created during their experiment conceptualization phase.

The goal of this tight envelope was to prevent combined inward and outward leakage between B2 and B1, in order to prevent B1 from contaminating B2, and ultimately to test the balance of internal B2 recycling flows.[11] As William Dempster, Project Engineer, explained,

> if the rate of leakage is small compared to the rates of gas exchange involved in the ecological processes, then the closed system will be a powerful instrument to study those processes . . .[12]

While their 100-year lifespan design specs for corrosion resistance and leak tightness challenged the engineering and construction industries, they knew that comparable construction tolerances had been specified for containment and fuel pools and nuclear power plant hot cell liners[13] and, hence, were possible. But their project required breakthrough innovations in air-tight technologies,[14] co-produced between project architects Margaret Augustine and Phil Hawes, Dempster and their cadres of scientist and engineering sub-contractors. Ultimately, B2's first two-year mission exchange rate was estimated, via two independent methods, at less than approximately 10 percent/year.[15]

Their initial calculations indicated that in order to achieve their target 10 percent annual leak rate, they could have no more than the equivalent of a 3/4-inch diameter penetration within the envelope. This total penetration area spanned across approximately 50 miles of caulked glazing joints and close to 10 miles of stainless-steel welds. Their post-experiment calculations concluded that even a leak rate of 400 percent/year (4 air-changes of 6,534,102 cubic feet/year) would have rendered their subtle atmospheric chemistry measurements illegible.[16] Every envelope penetration had to be scrutinized. As Dempster stated:

> There would be no aspect of the enclosure design that would permit flow of air between inside and outside. . . . For example, a stranded electrical wire to penetrate the enclosure could not be sealed only at its insulation jacket because air could flow between the strands inside the insulation.[17]

In addition, their envelope demanded other challenging design requirements: maximum admittance of sunlight for plant growth and photosynthesis, an impenetrable bathtub liner to prevent exchange with the ground below, airtight penetrations for utilities, and control of air pressure/volume expansion and contraction so as not to explode or implode the structure.[18] In addition, all joints needed to be permanently accessible for monitoring and repair.[19]

Consistent with the Biospherians' *synergetic* approach, they adopted the NASA Apollo missions accelerated development strategy of "'all-up systems testing' rather than exhaustive component by component analysis."[20] Their first attempt was the on-site Test Module, a *mini-B2* approximately 23 feet cubed, sealed utilizing two test methods, where they trialed the first application of an attached but distinct variable air-volume chamber endearingly referred to as *The Lung*. The Test Module was glazed on all sides, contained a small living area, lots of respiring plants, a vegetated *living machine* wastewater treatment system, an air handler and an analytic sensor system. It held at least four individual *test subject humans* for various durations totaling 60 days, the longest being three weeks, and achieved a leak rate of 24 percent/year.[21] Based on their Test Module enclosure experiments, the Biospherians were confident they could decrease B2's leak rate to within their target level.

No expense was spared to accomplish the less than 10 percent annual leak rate. Above ground, B2's space-frame envelope consisted of over 6,600 triangular panes of glass totaling 3.95 acres of glazing, requiring close to 50 miles of caulk. The east and west halves of B2 were connected with a flexible synthetic rubber fabric expansion joint, fitted with a leak detection system along its perimeter. Two (redundant) domed *Lungs* were attached to B2 via underground tunnels, and the *Energy Center* complex sat approximately 275 feet from B2, connected via underground pipes and wiring. All airlock doors were sealed with rubber gaskets wiped with silicone high vacuum grease. Those on the *outside* of B2's airlocks were fitted with pressure relief water traps topped with a floating oil layer to prevent evaporation. These traps allowed minimal air bubbles to escape only in the case of critical internal pressure levels and, with regular maintenance, ensured airlock integrity.[22]

Bathtub

The prevention of atmosphere, water, microbe and insect transfer between B2's foundation and the surrounding soil was paramount. Stainless steel was deemed a more durable liner than fiberglass or elastomerics, given the weight of the concrete structure and pressure of tons of soil in direct contact with the basement sidewalls.[23] A 1/8-inch thick[24] proprietary *super* stainless alloy, Allegheny Ludlum 6XN (20 Cr–25 Ni–6 Mo)[25] was developed by metallurgical engineer, Bob Walsh at Allegheny Ludlum.[26] This alloy proved more resistant to soil and seawater corrosion than other alloys. Approximately 500 tons of it lined the entire foundation and basement side walls of B2.[27]

While proving resistant to observable corrosion, Ludlum 6XN was not corrosion proof. Aquatic emersion tests of the stainless-steel liner samples were conducted at the Smithsonian Institution's recycling marine life system, which revealed rapid nickel, chromium and molybdenum water contamination in levels high enough to threaten marine life systems. Ultimately, this led to a final application of epoxy-based sealant over the entire liner surface. Cadmium fastener coatings and copper piping were excluded from B2 for the same reason.[28] In addition, Ludlum 6XN had the potential to corrode where it came into contact with the space-frame. Here, an air-tight, flexible seal of silicone elastomeric caulk mated a half-inch gap maintained between the liner, metal plates, and space-frame to prevent electrolytic corrosion.[29]

Securing the liner in place, and distributing column point loads, were necessary to ensure foundation integrity and the airtight seal. In order to sufficiently distribute structural column loads while protecting the bathtub from puncture, oversized double-anchored stainless-steel plates were welded and poured into the continuous concrete topping slab. The same construction detail was executed where beams were tied into concrete-lined side walls. In addition, structural tube-like anchoring members were welded and embedded within the concrete topper.[30]

In all, the stainless steel liner required almost 10 miles of welded seams,[31] while the concrete floor topping slab rendered some of these seams inaccessible to leak identification and maintenance (close to 2.5 miles).[32] To rectify this, the Biospherians designed a leak detection system of *air sniffer* tubes behind each weld, which fed into an approximately 6-foot diameter tunnel that lined the perimeter of B2's basement, underground and outside of the enclosure envelope.[33] The tunnel contained vacuum piping and telephone communication infrastructure which was used to test each weld and utility penetration for air tightness.[34]

Seal Tests

The liner seal tests consisted of four separate techniques: a weld test, a sniffer line/flood test, and two different positive pressure tests. During the weld test, a soap solution was painted over a 3-foot section of each weld, then a custom hand-held transparent plate (approximately 3.3 feet long, with a soft rubber gasket attached to its lower surface) was pressed over the top of the test seam. Next, a vacuum pump evacuated the air in the cavity created by the rubber gasket, between the plate and the seam.[35] Any gaps in the weld would cause bubbling. Once identified, the leak was repaired and an additional weld was applied to the seam. The new weld was retested, and repaired, until it no longer bubbled.

During the sniffer line/flood test, which occurred after the liner was welded but before the topping slab was poured, all flat basement surfaces were flooded with a few inches of water. Compressed air was forced through each individual sniffer line stub beneath the liner. Water bubbles revealed where gaps were present.[36]

Once B2 was enclosed, the most sensitive techniques, the positive pressure tests, were conducted. In the first test, a person inside B2 ran a canister of sulfur hexafluoride or SF_6 along each weld. Pinhole gaps in the weld seams would admit SF_6 through the envelope, where it was detected by sensors inside its corresponding *air sniffer tube*. The sensor sounded an alarm, which could be heard via a telecommunications link by the person inside conducting the test. As Dempster explained, "once detected, the pinhole could be precisely located, on the order of a centimeter, buy repeatedly positioning the gas release point and releasing tiny amounts while listening for a detector alarm."[37] During the first months of Mission 1, about a dozen pinhole leaks were identified and repaired using this technique.[38] The total area of leaks located in the liner closely approximated the total leak rate confirmed by another test technique, the *Lung Test*, described below.

In the second test, the *Lung Test*, again, harmless trace gas sulfur hexafluoride or SF_6 (heavy at mol. Wt. 146) in addition to helium (He, light at mol. Wt. 4) were introduced into B2's atmosphere. *Heavy* SF_6 primarily indicated bulk leakage in envelope penetrations, while *light* He indicated microscopic molecular diffusion through pathways or permeation through materials. B2's atmosphere was then pressurized utilizing the *Lungs* (as detailed later in the chapter), and air samples were taken from the *air sniffer tubes* and tested for the presence of these gasses, which indicated leaks. Ultimately, the dilution of both SF_6 and He were detected at similar rates.[39] As Dempster explains, "the implication is that there was very little leakage in the glazing."[40]

Space-frame

B2's space-frame glazing enclosure system was designed by Dempster and the engineering firm Pearce Structures. Founder Peter Pearce, an accomplished architect/engineer in his own right, had studied with Buckminster Fuller and worked with Charles and Rae Eames, among others.[41] The scope of the firm's work on B2 included not only space-frame strut and glazing design and testing to meet the challenging engineering requirements, but simultaneously addressing the Biospherian Architects' formal requirements, which included barrel vaults and other geometries inconsistent with inherent space-frame structural logics.

The double-layer space-frame was comprised of approximately 77,000 struts. Struts were almost 5 feet in external dimension and appeared identical, but were varied in their internal thickness depending on the specific structural stresses they needed to accommodate. They required unique numbering to ensure proper installation. They were fabricated from aluminum flame-sprayed coated steel tubing and electrostatically powder-coated white. Prior to installation, accelerated and concentrated salt water strut tests revealed excellent corrosion resistance results.[42]

In an attempt to mitigate complexity and potential construction errors, Pearce innovated the first integrated space-frame joint, the trademark "Multihinge Connection System."[43] This joint, which was fabricated as part of the strut, ensured fewer loose parts on site during construction as all attachments were accomplished with nuts, bolts and washers only, not a separate joint component.[44]

FIGURE 13.2 Pearce's Multihinge Connection System, which could accommodate multiple connection configurations.

Source: Photos by author.

Reconciling the structural forces of B2's space-frame was second only to the complications associated with airtight sealing of the space-frame's glazing. Ultimately, Dempster and Pearce developed a composite glass/plastic glazing system, where a quarter inch of PVB was heat-strengthened and laminated between two pieces of 1/4-inch glazing. This rendered the panels resistant to "ice balls fired by a pitching machine at approximately twice the velocity of hail" with a design wind load of approximately 2,400 Pa.[45]

To ensure consistent quality standards across B2's 6,600 triangular glass panes, they were factory produced: laminated, fitted, and caulked inside steel frames. Once vacuum tested for air tightness, panels were transported to site, bolted and sealed into the space-frame utilizing a two-stage caulking procedure. First, gray silicone caulk sealed the panel frame to the space-frame, then the process was repeated with white silicone caulk to ensure visual confirmation of a double-seal. As of 1999, five of B2's glass panels had cracked, but only on one side of their double-glazing lamination, and none compromised their airtight seal, because their plastic layer remained intact.[46]

With a glazing surface area of 3.9 acres[47] requiring almost 50 miles of caulk, the selection of a sealing medium was critical. Compared to other available elastic caulking mediums, silicone sealant had the longest estimated lifetime in reported field experience; was relatively resistant to UV, fatigue and creep; and was stable under significant temperature variation. In addition, it was materially non-organic, which was an important consideration for the prevention of insect damage.[48] Termites, which are voracious and versatile consumers, were a critical species in the mix of ecological decomposers within B2, but posed a potential threat to sealing mediums. Termite appetites for silicone were tested by sandwiching laboratory filter paper (a known termite food source) between two glass laboratory slides and carefully sealing them with silicone but leaving an exposed *tail* of filter paper as bait. "The test showed that the termites readily ate the exposed strip of paper but did not eat or chew through the caulking."[49] Ironically, the test termite species succumbed to extinction within B2, but a common and voracious Arizona greenhouse *crazy ant* (*Paratrechina longicornis*) proved problematic. Ultimately eight leaks, caused by crazy ants, had been identified as of 1999, easily located by following their trails.

Diurnal temperature differentials within B2, and temperature differentials across the envelope, required design consideration. In winter, significant interior condensation on the glazing was collected via specially designed troughs attached to the base of the glazing panes and routed to water tanks whence it was redistributed for irrigation, among other uses. On a sunny summer day, it was calculated that interior temperatures, without air conditioning, could reach 149°F, easily fatal to life inside.[50] The air volume expansions and contractions associated with these temperature, vapor and barometric pressure differentials were not trivial. To compound matters, the 541-foot straight space-frame span across the ocean and savannah biomes was structurally more sensitive to pressure differentials than a curved surface would have been. These "large pressure differences pose a danger that the structure will burst, either as an explosion (higher pressure inside) or an implosion (lower pressure inside)."[51]

The Lungs

B2's envelope was subject to projected atmospheric pressures on the order of ±5000 Pa (0.725 psi), far greater than the structure could withstand. Even if B2's skin was built strong enough to withstand these implosion and explosion forces, "the leak rate would be greatly increased as the substantial pressure variations alternately drove air in and out through any small holes."[52] In order to mitigate *pumping* air pressure fluctuations, caused by differential pressures between inside and outside, two variable volume chambers, or domed *Lungs*, were attached to the building. Essentially acting as eardrums by facilitating pressure equalization, the *Lungs* mitigated the *pumping* caused by combinations of three factors: temperature changes (over a range of 50° to 111°F,[53] which could expand or contract the interior atmosphere by ±3,500 Pa); humidity variations (1–100 percent, which could increase the air volume through vapor evaporation or contract it through condensation by ±500 Pa);[54] and typical Tucson external barometric pressure variations on the order of ±500 Pa.[55] In addition to volume-pressure compensation and maintaining the low leakage rate of under about 10 percent/year, "the lungs facilitated leakage measurement and detection."[56]

Each lung dome was comprised of an internal, flexible, air-tight membrane/lung pan apparatus and an external rigid weathercover dome, which operated in tandem with blowers, lung position sensors, barometric pressure sensors, temperature sensors and relative humidity sensors.[57] Each dome was approximately 160 feet in diameter and 24 feet high and at full capacity could contain 19,607 square feet of atmosphere, approximately 30 percent of B2's fixed air volume. The two lungs were each connected to B2 by ducts tall enough for a person to walk through, fitted with airlock doors and designed for independent and redundant operation.[58]

The air-tight flexible membrane/pan apparati (which operated similarly to an eardrum) consisted of a 93.5-foot diameter aluminum lung pan weighing 16 tons, seamlessly adhered to a 4-ton continuous flexible Hypalon membrane, whose shape and excess material facilitated expansion and contraction into any position within its range of motion.[59] Generally, the lung membrane either rested on the floor, when not in use and empty to limit (see *bottom* image of Figure 13.3), or was inflated, its lung pan/Hypalon membrane *ceiling* resting on top of the air in the lung chamber (see *top* image of Figure 13.3). The lung pan floated upward and fell "as B2's internal air volume fluctuated, propelling air as fast as 1.64 feet cubed/ second in or out of the lung."[60]

In case of excessive over-pressurization, each *Lung* had three emergency pressure relief valves in the form of gasket seals, bolted to thin glass panes within the

FIGURE 13.3   Interiors of the Lungs. *Top*: South Lung inflated and occupied under the membrane/pan apparatus. *Bottom*: West Lung deflated and occupied above the membrane/pan apparatus.

Source: Photos by author.

pan. "If a lung pan rose above its upper limit, a chain automatically pulled taut and a lever handle with a hammer head would break the thin glass pane, permitting excess air to vent outside."[61] Once the de-pressurized internal atmosphere returned the lung pan to its design-range of motion, the lid would fall back into position to halt excessive air loss. In order to re-establish the airtight seal, the broken glass pane required replacement.[62]

In cases of dramatic air-volume shifts, atmospheric pressure inside B2 was controlled by blowers between the weathercover and membrane/pan apparati. The blowers generated enough negative pressure (upward suction) to neutralize the weight of the lung pan.[63]

> This patented arrangement permit(ted) the air pressure within Biosphere 2 to be maintained anywhere from modestly positive (150 Pa, or 0.02 psi, when the membrane/lung pan apparatus is allowed to rest on the lung air column) to very slightly negative, relative to external barometric pressure.[64]

This level of control allowed B2 operators not only to limit leak rates but also to measure them.[65] "The lungs therefore provide for system-wide verification and precise measurement of the degree of sealing achieved,"[66] which confirmed a leak rate between 6.6 percent and 10 percent of total atmospheric volume, measured by the height of the lung pan.[67]

Quantifying Air Quality Within Tight Envelopes

B2's high degree of closure not only necessitated the exclusion of all off-gassing materials but also simultaneously facilitated the observation and measurement of carbon dioxide–oxygen exchange between ecosystems and atmosphere.[68] B2's Continuous Air Quality Monitoring System (CAQMS sensors) simultaneously facilitated the identification and maintenance of other atmospheric chemicals, including potentially toxic levels of nitrogen, ammonia, nitric oxide, nitrogen oxides, nitrous oxide (laughing gas), hydrogen sulfide, sulfur dioxide, carbon monoxide, methane and total non-methane hydrocarbons, from six locations.[69] As Dempster and project ecologist Mark Nelson explain:

> The initial two year closure in Biosphere 2 revealed the sharp fluctuations in atmospheric cycling that will be expected in small closed systems because of its concentration of living biomass and small air volumes.[70]

For example, B2's CAQMS system was so sensitive, it registered drops in photosynthesis carbon dioxide production as clouds passed overhead.

During the first two years of enclosure, carbon dioxide concentrations ranged from under 1,000 ppm to over 4,000 ppm (compared to Earth's 2018 average of 407 ppm).[71] While lessons were learned about high carbon dioxide atmospheric concentrations, such as magnified ocean acidification rates, it was accounting for B2's oxygen levels that proved the most surprising and challenging.

The lungs, in combination with the CAQMS system, facilitated the identification and measurement of significant losses of oxygen in B2's atmosphere over time. At the commencement of Mission 1, oxygen levels were at 20.9 percent, but by the sixteenth month of inhabitation, the Biospherians were living

> at 14.4% oxygen and total atmospheric pressure of 88 kPa . . . (1,160 m or 3,800 ft elevation) . . . the oxygen partial pressure was equivalent to Earth's atmosphere at 4,160 m or 13,650 ft elevation. This approaches the elevation of the highest human settlement.[72]

B2's atmosphere was losing oxygen, but "the obvious explanation of loss to (animal) respiration was not evident because the implied increase of carbon dioxide did not occur, and furthermore, the atmospheric volume of Biosphere 2 was shrinking"[73] with more than an unaccounted for 10 percent loss of atmospheric oxygen.[74] Where was the oxygen going?

After many months of investigation into the chemical concentrations in B2's soils, biomass, atmosphere, CO_2 scrubber product, and interior and exterior structural concrete, an

> isotopic analysis for carbon-12: carbon-13 ratios . . . revealed that the interior structural concrete of Biosphere 2 had sequestered substantial quantities

of CO_2 that roughly accounted for the *missing* amount by the reaction: $CO_2 + Ca(OH)_2$ ◊ $CaCO_3 + H_2O$.[75]

Approximately 26 tons of oxygen was absorbed over 170,070 square feet of exposed concrete surfaces.[76] This came as a significant finding, as carbonation of concrete within high humidity and concentrated carbon dioxide environments had not previously been studied. Comparative core analysis of concrete taken from B2's interior and exterior surfaces in 1994 and 1995 revealed rapid and advanced aging of interior surfaces, showing 10 to 25 mm carbonation depths.[77]

In January of 1993, after crew members showed signs of decreasing cognitive performance, a measured amount of pure oxygen was injected into B2's atmosphere to replace the oxygen in the carbonated molecules. Without B2's extremely tight envelope, which facilitated precise quantification of atmospheric chemical composition, it may not have been possible to identify the carbonation phenomenon and to locate concrete as the source of the oxygen's sequestration.

Enclosure Performance: Atmospheric Maintenance Regimes *in the Anthropocene*

As in B2, within B1's (Earth's) atmosphere, it has proven difficult to identify and quantify the increasing atmospheric carbon concentrations contributing to climate change. It has proven even more difficult to quantify additional potential carbon sources such as permafrost methane stores and possible carbon sequestration-sinks. As we increasingly conceptualize Earth's biogeochemistry as a Gaia-type closed-loop cybernetic system, where the outer reaches of our atmosphere perform as Earth's material envelope, we are confronted with questions about how Earth's closed-loop biogeochemical cycles operate, and at what rates. Simultaneously, we are confronted with our own cybernetic contribution to this evolving biogeochemistry.

B2's tight envelope was an architectural manifestation of the scientific concept of closure, which transformed the Biospherians, their technologies and their environment into a rapidly-cycling cyborgian system, not dissimilar to our current entangled Anthropocentric condition. In 1991, ironically in the same year that the Biospherians commenced Mission 1, Donna Haraway described these types of entanglements in *Simians, Cyborgs, and Women: The Reinvention of Nature*:

> A cyborg is a hybrid creature, composed of organism and machine. . . . Cyborgs are post Second World War hybrid entities made of, first, ourselves and other organic creatures in our unchosen "high-technological" guise . . . as communications systems, texts, and self-acting, ergonomically designed apparatuses.[78]

B2, as an eco-technical feedback-apparatus for human long-duration life-support, was clearly a manifestation of one of Haraway's cyborgs, a hybrid creature whose tight skin facilitated complex whole-system evolution, and its quantification, through atmospheric measurement.

Although B2 quickly diverged from Earth's biogeochemistry, its small size, relative to Earth's, provided some clues that may assist us in observing, conceptualizing, and testing how fast atmospheric greenhouse gas thermochemical reactions can occur. As Nelson and Dempster explain, "fluxes of biogeochemical elements can be rapid in small enclosed ecological systems because of the high concentrations of biotic elements, and small buffer capacities."[79] Re-examination of the *magnified* reactions and residence times within B2 (". . . the residence time for carbon dioxide in Earth's atmosphere is estimated at 3 years, in Biosphere 2 during its closure experiment, 1991–1994, carbon dioxide residence was approximately 4 days . . .")[80] might assist scientists in better calibrating climate change scale and timeframes, and may provide impetus and inspiration for novel carbon sequestration approaches and new means of detoxifying our bioaccumulating environment.

In addition, the Biospherians have described their most important lesson from B2 as one that linked scale and envelope-boundary to ethic, a type of *situated knowledge*,[81] coupled with appropriate behavior. They observe that Earth's biosphere is too large for one person to experience its boundary, or even to see ". . . the bounds of the biome or eco-region he or she is in . . ."[82] In contrast,

> Biosphere 2 . . . put(s) the boundaries of the whole and the relation of the parts into a clearly visible paradigm or model that hopefully delineates the ethics and necessity of an ecological lifestyle for inhabitants of a biosphere.[83]

B2 produced environmental discoveries increasingly relevant to our current conceptualizations and observations of Earth's climate via cyborgian formations, which redefined and blurred boundaries between bodies, technologies and natures. In an ironic inversion of James Marston Fitch's functionalist post-World War II framing of peoples' connection to their environment through the architectural envelope's technological filtration, the Biospherians, as cyborgs, experienced *the load of the "natural" environment* as a self-produced *Cabin Ecology*, generated via their envelope's construction and technologies. B2's envelope, and the eco-technical assemblage of humans and non-humans it contained, operated as a tool of knowledge production, allowing the Biospherians to produce truth-claims and life-ways tuned to an anthropocentrically and synthetically constructed world. Their experiment provides useful understandings today, as envelopes are engineered tighter to increase energy efficiency with dangerously diminished concern for atmospheric chemistry and air-change, as we pursue ever more ambitious *Cabin Ecology* reliant explorations into non-Earth environments, and ultimately, to better understand how our enclosed, dynamic, cyborg and increasingly Anthropocentrically driven planet is changing.

Notes

1. Martica A. Sawin, ed., *James Marston Fitch: Selected Writings on Architecture, Preservation, and the Built Environment* (New York and London: W. W. Norton & Company, 2006), 15.
2. James Marston Fitch, *American Building: The Forces That Shape It* (Boston: Houghton Mifflin Co., 1948), 146.
3. Ibid., 148.
4. Branden Hookway, "Cockpit," in *Cold War Hothouses: Inventing Postwar Culture, From Cockpit to Playboy*, eds. Beatriz Colomina, Annmarie Brennan and Jeannie Kim (New York: Princeton Architectural Press, 2004), 42.
5. W.B. Cassidy, ed., *Bioengineering and Cabin Ecology* (Tarzana: American Astronautical Society, Science and Technology Series, Volume 20, 1968).
6. Bernard Zabel, Phil Hawes, Stuart Hewitt and Bruno D.V. Marino, "Construction and Engineering of a Created Environment: Overview of the Biosphere 2 Closed System," *Ecological Engineering* 13 (1999): 47.
7. William F. Dempster, "Methods for Measurement and Control of Leakage in CELSS and Their Application and Performance in the Biosphere 2 Facility," *Advances in Space Research* 14, no. 11 (1994): 331.
8. John Allen, *Biosphere 2: The Human Experiment* (New York: Penguin Press, 1991), 156 pages.
9. Bruno Latour, *We Have Never Been Modern* (Cambridge: Harvard University Press, 1991), 142–45.
10. William F. Dempster, Mark Nelson and John Allen, "Achieving Closure for Bioregenerative Life Support Systems: Engineering and Ecological Challenges, Research Opportunities" (conference paper, F4.5 Session, COSPAR, Bremen, Germany, 2010), 2.
11. Ibid., 3.
12. William F. Dempster, "Biosphere 2 Engineering Design," *Ecological Engineering* 13 (1999): 32–33.
13. Zabel, Hawes, Hewitt and Marino, "Construction and Engineering of a Created Environment," 47.
14. Mark Nelson and Willian F. Dempster, "Living in Space: Results from Biosphere 2's Initial Closure, an Early Testbed for Closed Ecological Systems on Mars," in *Strategies for Mars: A Guide to Human Exploration*, eds. Carol R. Stoker and Carter Emmart (San Diego: Univelt, Inc., Science and Technology Series 86, 1996), 368.
15. Dempster, "Biosphere 2 Engineering Design," 33.
16. William F. Dempster, "Tightly Closed Ecological Systems Reveal Atmospheric Subtleties—Experience from Biosphere 2," *Advances in Space Research* 42 (2008): 1952–53.
17. Ibid., 1952.
18. Dempster, "Biosphere 2 Engineering Design," 33.
19. Zabel, Hawes, Hewitt and Marino, "Construction and Engineering of a Created Environment," 46.
20. Nelson and Dempster, "Living in Space: Results From Biosphere 2's Initial Closure," 368.
21. Ibid., 368–69.
22. Zabel, Hawes, Hewitt and Marino, "Construction and Engineering of a Created Environment," 54.
23. Dempster, "Biosphere 2 Engineering Design," 34.
24. Dempster, Nelson and Allen, "Achieving Closure for Bioregenerative Life Support Systems," 5.
25. William F. Dempster, "Biosphere II: Closed Ecological Systems Engineering," *Engineering, Construction, and Operations in Space II: Proceedings of Space* 90, no. 2 (1990): 1212.
26. Allen, *Biosphere 2: The Human Experiment*, 63–64.
27. Dempster, "Biosphere 2 Engineering Design," 34.

28. Dempster, "Biosphere II: Closed Ecological Systems Engineering," 1212.
29. Zabel, Hawes, Hewitt and Marino, "Construction and Engineering of a Created Environment," 47.
30. Ibid., 47.
31. Dempster, "Tightly Closed Ecological Systems Reveal Atmospheric Subtleties," 1952.
32. William F. Dempster, "Airtight Sealing a Mars Base," *Life Support & Biospheric Science* 8 (2002): 156.
33. Zabel, Hawes, Hewitt and Marino, "Construction and Engineering of a Created Environment," 47.
34. Dempster, Nelson and Allen, "Achieving Closure for Bioregenerative Life Support Systems," 5.
35. Zabel, Hawes, Hewitt and Marino, "Construction and Engineering of a Created Environment," 48–49.
36. Dempster, Nelson and Allen, "Achieving Closure for Bioregenerative Life Support Systems," 5–6.
37. Zabel, Hawes, Hewitt and Marino, "Construction and Engineering of a Created Environment," 48–49.
38. Dempster, Nelson and Allen, "Achieving Closure for Bioregenerative Life Support Systems," 5–6.
39. Nelson and Dempster, "Living in Space: Results From Biosphere 2's Initial Closure," 384–85.
40. Dempster, Nelson and Allen, "Achieving Closure for Bioregenerative Life Support Systems," 6.
41. Allen, *Biosphere 2: The Human Experiment*, 62.
42. Zabel, Hawes, Hewitt and Marino, "Construction and Engineering of a Created Environment," 49.
43. Allen, *Biosphere 2: The Human Experiment*, 62.
44. Ibid., 62.
45. Ibid., 64.
46. Zabel, Hawes, Hewitt and Marino, "Construction and Engineering of a Created Environment," 51.
47. Nelson and Dempster, "Living in Space: Results From Biosphere 2's Initial Closure," 375.
48. Zabel, Hawes, Hewitt and Marino, "Construction and Engineering of a Created Environment," 51–52.
49. Ibid., 51–52.
50. Dempster, "Biosphere 2 Engineering Design," 37, 39–40.
51. Dempster, Nelson and Allen, "Achieving Closure for Bioregenerative Life Support Systems," 3.
52. Dempster, "Biosphere 2 Engineering Design," 35.
53. Zabel, Hawes, Hewitt and Marino, "Construction and Engineering of a Created Environment," 52.
54. Dempster, "Tightly Closed Ecological Systems Reveal Atmospheric Subtleties," 1954.
55. Dempster, "Methods for Measurement and Control of Leakage," 331.
56. Dempster, "Biosphere 2 Engineering Design," 31.
57. Dempster, "Methods for Measurement and Control of Leakage," 333.
58. Dempster, "Biosphere II: Closed Ecological Systems Engineering," 1211.
59. Dempster, Nelson and Allen, "Achieving Closure for Bioregenerative Life Support Systems," 4.
60. Zabel, Hawes, Hewitt and Marino, "Construction and Engineering of a Created Environment," 52–53.
61. Ibid., 53.
62. Ibid., 53.
63. Dempster, "Biosphere 2 Engineering Design," 35.

64. Nelson and Dempster, "Living in Space: Results From Biosphere 2's Initial Closure," 375.
65. Ibid., 375.
66. Ibid., 375.
67. Dempster, "Biosphere II: Closed Ecological Systems Engineering," 1211.
68. Zabel, Hawes, Hewitt and Marino, "Construction and Engineering of a Created Environment," 53.
69. Dempster, "Biosphere 2 Engineering Design," 31.
70. John Allen and Mark Nelson, "Overview and Design: Biospherics and Biosphere 2, Mission One (1991–1993)," *Ecological Engineering* 13 (1999): 25.
71. Nelson and Dempster, "Living in Space: Results From Biosphere 2's Initial Closure," 363.
72. Ibid., 363.
73. Dempster, "Airtight Sealing a Mars Base," 157.
74. Dempster, "Biosphere 2 Engineering Design," 35.
75. Nelson and Dempster, "Living in Space: Results From Biosphere 2's Initial Closure," 386.
76. Ibid., 386.
77. Dempster, "Biosphere 2 Engineering Design," 37.
78. Zabel, Hawes, Hewitt and Marino, "Construction and Engineering of a Created Environment," 54–55.
79. Donna Haraway, *Simians, Cyborgs, and Women: The Reinvention of Nature* (New York: Routledge, 1991), 1.
80. Nelson and Dempster, "Living in Space: Results From Biosphere 2's Initial Closure," 381.
81. Dempster, Nelson and Allen, "Achieving Closure for Bioregenerative Life Support Systems," 8.
82. Donna Haraway, "Situated Knowledges: The Science Question in Feminism and the Privilege of Partial Knowledge," *Feminist Studies* 14, no. 3 (1998): 575–99.
83. Mark Nelson, Philip Hawes and Kathleen Dyhr, "Biosphere 2: Laboratory, Architecture, Paradigm & Symbol," *IS Journal* 10 (1990): 66.

INDEX

Note: Page numbers in *italic* indicate a figure on the corresponding page.